Passive Optical Components for Optical Fiber Transmission

For a complete listing of the *Artech House Antenna Library*,
turn to the back of this book

Passive Optical Components for Optical Fiber Transmission

Norio Kashima

Artech House
Boston • London

Library of Congress Cataloging-in-Publication Data
Kashima, Norio.
Passive optical components for optical fiber transmission / Norio Kashima.
Includes bibliographical references and index.
ISBN 0-89006-775-9 (hard)
1. Optical communications–Equipment and supplies. 2. Optical fibers. 3. Optical fibers–Joints.
  I. Title.
TK5103.59.K38   1995
621.382'75–dc20

94-39307
CIP

British Library Cataloguing in Publication Data
Kashima, Norio
Passive Optical Components for Optical Fiber Transmission
I. Title
621.38275

ISBN 0-89006-775-9

© 1995 ARTECH HOUSE, INC.
685 Canton Street
Norwood, MA 02062

International Standard Book Number: 0-89006-775-9
Library of Congress Catalog Card Number: 94-39307

10 9 8 7 6 5 4 3 2 1

To my father and mother

Contents

Preface

Optical fibers with a loss of 20 dB/km were fabricated in 1970. Since then, optical transmission systems have become a reality, not a dream. Optical transmission systems for trunk lines using optical fibers are now used worldwide. In the future, these systems will penetrate deeply into the subscriber loops and local area networks. Although the key components for optical-fiber transmission systems are optical fibers and laser diodes, passive optical components such as optical connectors and optical couplers are also very important for constructing the systems. This book deals with passive optical components for optical-fiber transmission systems. Although it covers the passive optical components used for both single-mode and multimode fibers, those for single-mode fibers are emphasized. The technologies of passive optical components are explained basically and easily by considering their theoretical and practical aspects. This book is intended for field engineers, system designers, technical managers, students, and manufacturers who are interested in passive optical components such as fiber connections, optical filters, couplers, and switches. It is also useful for researchers who are beginning to investigate these fields.

This book is divided into three parts: common basic technologies for passive optical components in Part I, and individual technologies in Parts II and III. In Part I (Chapters 1 to 4), optical-fiber transmission systems, characteristics of optical fibers, coupling loss, reflection properties, and measurement methods are explained. In Part II (Chapters 5 to 9), simple connection components such as splices and connectors are treated. These were the major passive components in the early stage of optical-fiber transmission, and are inevitable for current and future optical-fiber transmission systems. Splice and connector technologies are intensively treated in this book because of their importance in realizing low-loss transmission lines. Without low-loss splice and connector technologies, low-loss transmission lines could not be realized even by using ultra-low-loss fibers. In this sense, simple connection components may be the third key components, next to optical fibers and laser diodes, for optical-fiber transmission. In Part III (Chapters 10 to 12), components with some functions, such as coupler, filter, and switch, are treated. These are not always used in the usual fiber transmission systems. However, they are very important for constructing a specific system, such as wavelength division multiplexing (WDM) systems.

Chapters 1 and 2 contain introductory material on passive optical components. Chapter 1 explains optical-fiber transmission systems where passive optical components are used, and the requirements for the components of these systems are explained. This chapter also contains a brief explanation of optical fibers and cables. Chapter 2 discusses the characteristics of fibers. Chapters 3 and 4 contain basic material on passive optical components. Chapter 3 treats coupling loss and mode conversion. Mode conversion between guided modes is unique to multimode fibers. The influence of mode conversion on baseband response is also explained. Chapter 4 deals with measurement methods for loss, reflection, fiber endface quality, and crosstalk. Chapters 5 and 6 discuss fusion splicing by electric discharge, gas-laser, and flame. Several important techniques are explained, such as the prefusion method, the high-frequency discharge with high-voltage trigger method, the self-alignment method, and the uniform heating method for mass-fusion splicing. Chapter 7 discusses both mechanical and adhesive-bonded splices. Chapter 8 deals with optical connectors: not only ordinary single-fiber connectors, but also array-fiber connectors and two-dimensional-fiber connectors. Chapter 9 discusses the splicing of special fibers, such as polarization-maintaining fibers, fluoride fibers, and erbium-doped fibers. These fibers are not widely used today, but they have attractive characteristics for special applications when compared to the ordinary silica-based single-mode or multimode fibers. Chapter 10 discusses directional and star couplers. Directional couplers are often used for bidirectional transmission over one fiber for the purpose of mixing or branching optical signals (lightwaves), and for many measurements. Star couplers are used for many optical signals (lightwaves) to mix or branch. They are often used for local-area network and video distribution systems. Chapter 11 explains WDM and optical frequency division multiplexing (OFDM) filters, which are used in WDM or OFDM systems. In addition to WDM and OFDM filters, several optical filters are also treated in this chapter. Chapter 12 discusses optical attenuators, isolators, and circulators, which are used for improving the system performance or functionality. Chapter 13 discusses the optical mechanical switches that apply fiber connection technologies. These are used in some transmission and measurement systems. Future technology from the author's viewpoint is discussed in Chapter 14.

I wish to thank my many colleagues at Nippon Telegraph and Telephone (NTT) for coresearching and codeveloping passive optical components. I also thank Mr. Mark Walsh for inviting me to write this book.

PART I
BASIC TECHNOLOGIES

Chapter 1

Introduction to Passive Optical Components

There are many passive optical components that have been or will be used for optical-fiber transmission. In this chapter, optical-fiber transmission systems, such as trunk, subscriber loop, and local-area network (LAN) systems, are explained. Several passive optical components are used in these systems. Optical fiber and cable are briefly explained here, and precise explanations of fiber characteristics will be provided in Chapter 2. The classification of passive optical components and requirements for fiber transmission are also explained in this chapter.

1.1 OPTICAL-FIBER TRANSMISSION SYSTEMS

1.1.1 History of Fiber Transmission

The research of dielectric waveguides has a long history, starting in the 1930s. The dielectric rod waveguide and the dielectric surface waveguide were proposed and investigated. Optical fiber is one of the dielectric waveguides. The research of optical fiber as a transmission line is a relatively new field, having started in the 1960s. In 1960, a ruby laser was developed by T.H. Maiman [1]. Along with the investigation of laser diodes, low-loss optical fiber was anticipated by K. C. Kao and B. A. Hockham in 1966 [2]. In 1970, continuous-wave lasing of a laser diode at room temperature was realized [3], and fused silica fibers with 20-dB/km loss were developed [4]. The realization of these key devices turned the dream of optical-fiber transmission into reality. Laser diodes with 0.8-, 1.3-, and 1.5-μm emitting wavelengths are currently available for optical transmission. Fiber loss has decreased to below 0.2 dB/km at a 1.55-μm wavelength at present [5]. When compared to the conventional metallic waveguide or cables, optical fibers have the following several advantages: (1) low loss (long-distance transmission), (2) broadband (high-speed transmission), and (3) no crosstalk and no electromagnetic interference (high-quality transmission). Using these advantages, several optical transmission systems have been investigated and used.

Telecommunication is classified into international and domestic communications. Domestic telecommunication is composed of trunk, subscriber loop, and LAN systems. The trunk system connects between telephone offices, and the subscriber loop system connects between a telephone office and subscribers (users). The LAN system is a localized system used only for the closed area. When the LAN is connected to public networks, the terminals of the LAN can communicate with other persons or computers located at a long distance away.

Historically, optical trunk transmission systems were the first to be investigated and introduced into real use. In the early systems, 0.8-μm-wavelength light and step-index (SI) multimode fibers were used. Next, 1.3-μm-wavelength light and graded-index (GI) multimode or single-mode fibers were used. Today's trunk systems generally use 1.3- or 1.5-μm-wavelength light, single-mode fibers, and distributed feedback laser diodes (DFB LD). Higher speed transmission systems including optical soliton transmission has been investigated for next-generation trunk systems. Construction of optical subscriber systems and optical LAN systems, the next major targets, have been intensively investigated. In the early investigation stage, optical subscriber loop systems using GI multimode fiber were taken into consideration. At present, single-mode optical-fiber systems are being investigated and have been considered to be the preferable systems. There are many passive optical components used for these fiber transmission systems. Among them, the optical splices and connectors are indispensable components for all systems. Shortly after the development of 20-dB/km loss fibers, the research of splices and connectors started. In Figure 1.1, the history of fiber transmission and related components is shown as a summary of this section.

1.1.2 Optical-Fiber Transmission

There are many digital and analog transmission technologies used for optical-fiber transmission [6]. These technologies can be categorized according to intensity modulation/direct detection (IM/DD) technology and coherent transmission technology. The principles of these technologies are shown in Figure 1.2. IM/DD technology is simple and has been used for many existing optical systems. In IM/DD, the intensity of laser light is either directly modulated or externally modulated by digital or analog electric signals (only direct modulation of a laser diode is shown in this figure). As a result of the modulation, light intensity varies in digital or analog format. The intensity-modulated light through a fiber is directly detected either by a PIN photodiode (PIN-PD) or by an avalanche photodiode (APD).

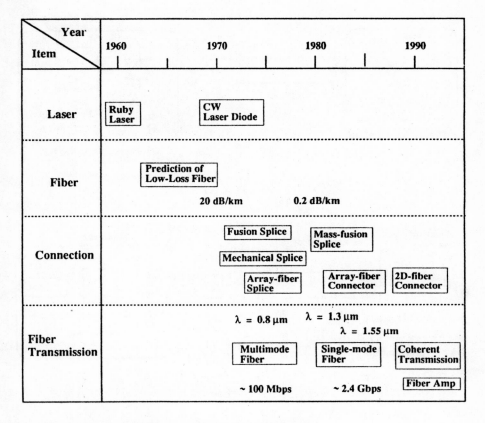

Figure 1.1 History of optical-fiber transmission and components.

Coherent transmission technology positively uses the wave aspect of light. The existing modulation and detection technologies, such as frequency-shift keying (FSK) modulation, phase-shift keying (PSK) modulation, heterodyne detection, and homodyne detection, can be applied for optical coherent transmission technologies. The principle of optical coherent transmission is similar to that used in the existing radio wave transmission. The received signals are mixed with the waves from the local oscillator in a receiver. Two lightwaves, signal and local oscillator waves, are mixed in coherently. In homodyne detection, both the optical frequency and phase of the local oscillator wave are identical and locked to those of the input signal wave. In heterodyne detection, the frequency of the local oscillator wave is different from that of the input signal wave. Signals are detected synchronously or asynchronously from

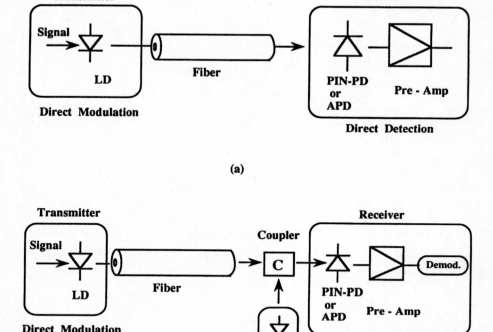

Figure 1.2 Principle of fiber transmission: (a) principle of IM/DD; (b) principle of coherent transmission.

the mixed waves in electrical processing. In optical coherent transmission, a photodiode (or an APD) is used as both an optical detection device and a mixing device. Coherent transmission technology has the advantages of higher receiver sensitivity and superior optical frequency selectivity over IM/DD. Higher receiver sensitivity makes it possible to realize a longer repeaterless transmission length. Using superior optical frequency selectivity, densely spaced wavelength-division multiplexing (WDM) is realized by optical coherent technology, known as optical frequency-division multiplexing (OFDM). The wavelength spacing of ordinary WDM is about 100 to 300 nm and that of OFDM is about 0.04 nm when the optical frequency space Δf is 5 GHz. OFDM using IM/DD is also possible with an OFDM filter. An OFDM filter commonly uses the principle of interferometers, such as the Mach-Zehnder (MZ) interferometer and the Fabry-Perot (FP) interferometer.

1.1.3 Optical Trunk System

A typical optical trunk system using fiber transmission is shown in Figure 1.3. The gathered information from subscribers is multiplexed in a telephone office. The multiplexed information is transmitted between telephone offices at a higher bit rate (in the case of digital transmission) using repeaters. At present, digital transmission using single-mode fibers is commonly used. In a repeater, received digitally modulated light signals are converted to electric signals by a photodiode and the converted signals are electrically reshaped through the so-called 3R functions (retiming, reshaping, and regenerating). And again the electric signals are converted to light signals by a laser diode. At present, 3R functions and amplification are performed by electric devices. Optical signal processing without converting to electric signal forms has been investigated for higher speed transmission. Optical amplification using a laser diode or erbium-doped fiber (EDF) has been realized at a laboratory and practically used in some systems. A typical repeater interval is from 50 to 100 km. Both fiber loss and dispersion influence the repeater interval.

1.1.4 Optical Subscriber Loop System

There are several optical subscriber loop systems. Some are actually in use and some are only at the trial or laboratory level. Figure 1.4 illustrates two types of systems: one is the direct connection from telephone office to subscribers using only optical cables, and another is the hybrid connection using optical and metallic or coaxial cables. In the latter case, optical-electrical conversion is made in a remote terminal (RT). The typical transmission length for a subscriber loop system is within 10 km.

1.1.5 Optical Cable Television System

The cable television (CATV) system provides many video channels to subscribers. Two possible network constructions are shown in Figure 1.5. The introduction of optical cables into this system has been made and the hybrid network construction of optical cables and coaxial cables are used at present. In this figure, the star coupler is used for distribution of video signals in the case of using all-optical cables. The typical transmission length for an optical CATV system is similar to that of subscriber loop systems.

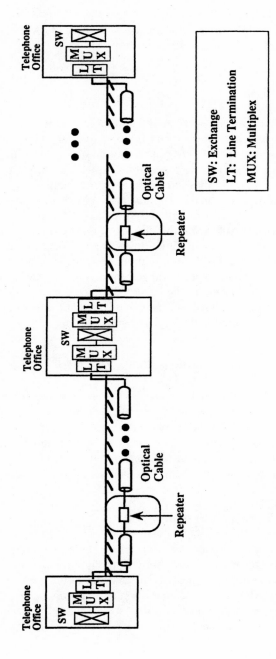

Figure 1.3 Optical trunk system.

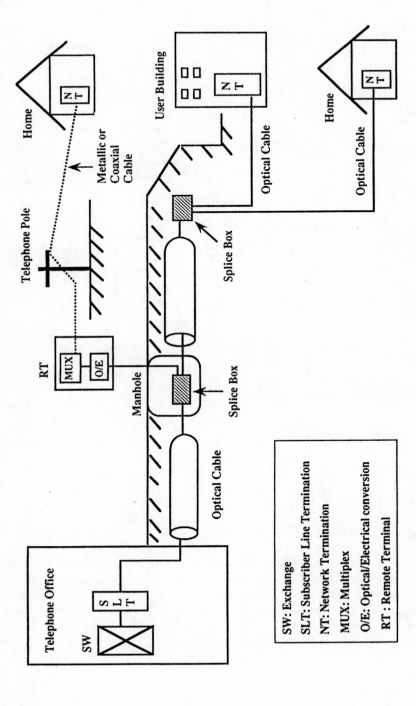

Figure 1.4 Optical subscriber loop systems.

SW: Exchange
SLT: Subscriber Line Termination
NT: Network Termination
MUX: Multiplex
O/E: Optical/Electrical conversion
RT : Remote Terminal

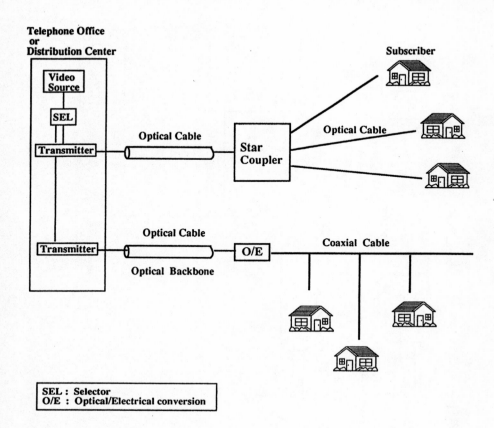

Figure 1.5 Optical CATV system.

1.1.6 Optical LAN System

A LAN is used for communication between terminals in a building or in a few neighboring buildings, and its typical transmission speed is around 1 to 100 Mbps [7]. Computers or work stations can be connected by using a LAN. In a fiber distributed data interface (FDDI), the transmission speed is 100 Mbps. Figure 1.6 shows an example of an optical LAN in a building used as an optical backbone. A LAN can be connected to public networks through a gateway.

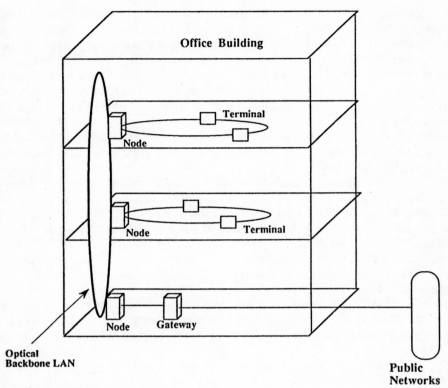

Figure 1.6 Optical LAN system.

1.2 OPTICAL FIBER AND CABLE

1.2.1 Classification of Optical Fiber

Light from an optical light source, such as a laser diode, propagates in optical fiber, which is the widest bandwidth transmission media at present. The materials making up optical fibers can be divided into five categories:

1. All-fused-silica-glass (SiO_2) fiber (silica-core with silica-cladding fiber);
2. Silica-core with plastic-cladding fiber;
3. Compound glass fiber such as fluoride glass fiber;
4. Plastic fiber;
5. Liquid-core fiber.

Among the above items, all-fused-silica-glass fiber has the lowest loss and superior reliability at present. Therefore, fused-silica-glass fiber is widely used for fiber transmission. Liquid-core fiber was investigated at the early stage of fiber investigation. As liquid is filled in a core, it is not easy to handle it. Therefore, it is not of practical use. It may be useful for the application of the optical nonlinear effect using a liquid. Both silica-core with plastic-cladding fiber and plastic fiber are easy to handle; however, they have larger loss when compared to all-silica fibers. Compound glass fiber was the alternative fiber at the early stage of fiber investigation. However, its transmission loss could not be lowered when compared to all-silica fibers. Recently, it has been anticipated that compound glass fiber such as fluoride glass fiber has the potential for ultra-low transmission loss (approximately 0.01 or 0.001 dB/km). Although intensive research has been done, ultra-low-loss compound glass fibers are not realized at present. In this book, all-silica fibers, including dispersion-shifted fiber (DSF) and EDF, and fluoride glass fibers are treated. For the sake of simplicity, the word fiber in this book means an all-silica fiber, and the term fluoride fiber is used to indicate a fluoride glass fiber.

Silica fiber is made from SiO_2. The refractive index of SiO_2 is about 1.45 to 1.46 near a 1-μm wavelength and its value slightly decreases for a longer wavelength. The basic phenomenon for a guiding light in a fiber is the total internal reflection, and light is confined in higher refractive-index materials. For a guiding light, the fiber cross section is not uniformly made. It has a slightly higher refractive-index part in the center, called the core. The core is surrounded by cladding as shown in Figure 1.7.

To form a core and cladding structure, several dopants are used to change the refractive index. For example, GeO_2 is used as a dopant for increasing the refractive index and fluorine (F) is for decreasing it. One example of manufacturing fiber is the combination of GeO_2 + SiO_2 for core and SiO_2 for cladding. By controlling the volume of GeO_2, the refractive-index profile of core $n_1(r)$ can be arbitrarily made, where r is the radial coordinate. Another example of manufacturing fiber is the combination of SiO_2 for core and SiO_2 + F for cladding. In this case, the refractive-index profile of core $n_1(r)$ is uniform. Generally, a fiber with a uniform core index is called a step-index fiber. The definition of an SI fiber is

$$n_1(r) = n_0 \qquad (0 \le r \le a) \qquad (1.1)$$

$$n_2(r) = n_2 \qquad (a \le r \le b) \qquad (1.2)$$

where n_0 and n_2 are constants and a is the core radius ($2a$ is the core diameter), and n_0 and n_2 are the refractive indexes of core and cladding, respectively. Fibers with a small refractive-index difference between core and cladding are usually called weakly guiding fibers [8]. All fibers used for optical transmission at present are weakly guiding fibers. The refractive-index profile for weakly guiding fibers is usually expressed using a parameter as follows:

Cross Section

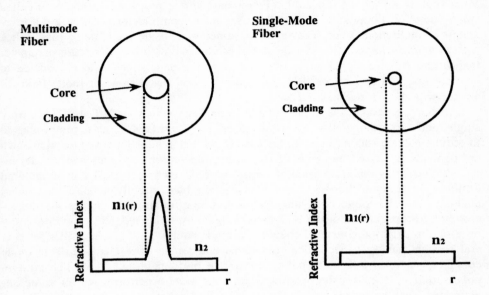

Multimode Fiber

Single-Mode Fiber

Core

Core

Cladding

Cladding

Refractive Index

$n_1(r)$

n_2

Refractive Index

$n_1(r)$

n_2

r

r

Figure 1.7 Structure of optical fiber.

$$n_1(r) = n_0[1 - 2\Delta(r/a)^{\alpha}]^{1/2} \tag{1.3}$$

$$n_2(r) = n_2 \tag{1.4}$$

where

$$\Delta = (n_0 - n_2)/n_0 \tag{1.5}$$

Since the fibers considered are weakly guiding fibers, $\Delta \ll 1$ holds. For SI fibers, $\alpha \to \infty$ (α tends to infinitive) in (1.3). The following normalized frequency parameter V is defined and is very useful.

$$V = 2\pi a(n_0^2 - n_2^2)^{1/2}/\lambda \tag{1.6}$$

$$\approx 2\pi a n_0 (2\Delta)^{1/2}/\lambda \tag{1.7}$$

where λ is the wavelength in free space. Equation (1.7) holds because of weakly guiding fibers.

For SI fibers, only one mode can propagate in a fiber when $V < 2.405$. The value 2.405 is called the cutoff value V_C. The mode is called the HE_{11} mode (or the

LP$_{01}$ (linearly polarized) mode as a weakly guiding fiber). For the general α value, V_C is slightly different from 2.405. Fibers having one propagating mode are called single-mode or monomode fibers. There are many combinations of a and Δ for realizing a single-mode fiber. They are determined by considering bending loss, splice loss, manufacturing conditions, and so on. Mode field diameter is a better parameter than a and Δ for a single-mode fiber, because the parameter is closely related to bending loss and splice loss [9]. The definition and discussions of mode field diameter will appear in Chapter 2.

Strictly speaking, the mode number in ordinary single-mode fibers is not truly single, and two orthogonally polarized modes, HE$_{11}^x$ and HE$_{11}^y$ modes, can propagate in ordinary single-mode fibers. In the case of perfect symmetric structure, these two orthogonally polarized modes have the same propagation constants and are degenerated. Perfect symmetric structure cannot be realized in practically manufactured ordinary single-mode fibers. Manufactured fibers have slightly noncircular core or cladding, or some other nonsymmetry. Nonsymmetric structure breaks the degeneration of propagation constants. Two orthogonally polarized modes with slightly different propagation constants couple (the mode coupling effect) in a fiber for several reasons, such as mechanical vibrations and temperature changes. This mode coupling generates a random polarization state of fiber output light. The random polarization state causes the deterioration of the system performance for some applications, such as coherent fiber transmission and coherent sensing. Although several countermeasures are proposed for suppressing the random polarization coupling, one basic solution is to use a polarization-maintaining fiber. There are many proposed polarization-maintaining fibers whose cross sections are shown in Figure 1.8 [10–16]. They are an elliptic-core fiber, an elliptic-jacket fiber, a polarization-maintaining and absorption-reducing (PANDA) fiber, a bow-tie fiber, and a side-pit or a side-tunnel fiber. These polarization-maintaining fibers are realized by using high birefringence generated by a nonsymmetric refractive-index profile in a core or near a core region. A nonsymmetric refractive-index profile is introduced by the nonsymmetric structure effect or by the nonsymmetric stress effect. Stress induces small refractive-index deviation (the photoelastic effect) [17]. In actual polarization-maintaining fibers, both effects of nonsymmetric structure and nonsymmetric stress exist. Usually, one effect is dominantly used to realize high birefringence in a polarization-maintaining fiber. The typical polarization-maintaining fiber using a nonsymmetric structure is an elliptic-core fiber, and that using nonsymmetric stress is a PANDA fiber. In a PANDA fiber, stress is applied by the boron-doped parts, which are located near the core and are shown as the shaded circles in Figure 1.8(d). Polarization-maintaining fibers have relatively large transmission loss when compared to ordinary single-mode fibers. Research to lower the loss has been done, and one successful example is the PANDA fiber [13].

Fibers with a large V value are called multimode fibers, since many modes propagate in these fibers. Multimode fibers with $\alpha = 2$ in (1.3) have a wider band-

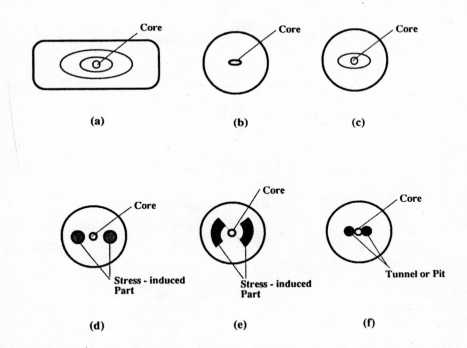

Figure 1.8 Structure of polarization-maintaining fibers: (a) noncircular cladding; (b) elliptic core; (c) elliptic jacket; (d) PANDA; (e) bow tie; (f) side tunnel or side pit.

width than fibers with other α values. They are called graded-index multimode fibers. The GI multimode fiber, rather than the SI multimode fiber, is usually used for fiber transmission because of its wider bandwidth. The characteristics of multimode fibers are complex because of the coupling effect of many propagating modes (the mode coupling effect).

Typical values of commonly used single-mode and multimode fibers are listed in Table 1.1. Because of the large core in a multimode fiber, this fiber has good coupling efficiency between a fiber and a laser and between fibers. A multimode fiber is used for relatively low speed fiber transmission, as indicated in Table 1.1. Many optical LAN systems use multimode fibers. Single-mode fibers have wide bandwidth, but they are difficult to splice. As the fiber splicing and connecting techniques in the field improve, single-mode fibers are commonly used for fiber transmission. A polarization-maintaining fiber is used only for special applications, such as passive components, and is not currently used for transmission lines. One major reason is that, with current techniques, polarization-maintaining fibers are difficult to splice or connect due to the fiber polarization axis.

Table 1.1
Multimode and Single-Mode Fibers*

		Multimode Fiber		Single-Mode Fiber	
		Step Index	Graded Index	Ordinary	PANDA
Core diameter	(μm)	30–80 (50)	30–80 (50)	8–10	8–10
Cladding diameter	(μm)	100–150 (125)	100–150 (125)	100–150 (125)	125–150
Δ	(%)	0.7–1.0 (0.7)	0.8–1.2 (1.0)	0.2–0.5	0.3–0.4
Loss	$\lambda = 0.8 \ \mu$m	2–3	2–3	2–3	2–3
(dB/km)	$\lambda = 1.3/1.5 \ \mu$m	0.3–1.0	0.3–1.0	0.2–1.0	0.3–1.0
Bandwidth (MHz km)		20–50	100–2,000	>1 GHz	>1 GHz

*Parentheses indicate typical value.

1.2.2 Fiber Fabrication Method

Fibers are made by drawing a fiber mother rod, which is called a preform. The typical diameter of a preform is a few centimeters. It is heated up to about 2,000°C by a furnace in the case of silica fiber drawing. The fiber drawing process is very important, because it determines the dimensional and mechanical properties of fibers. Fiber dimensions influence both optical transmission characteristics and fiber connecting characteristics, such as splice loss, which will be discussed in later chapters. During manufacturing, the fiber diameter is feedback-controlled by varying the drawing speed using fiber diameter monitoring signals. The monitoring signals are obtained usually by a contactless measuring method using a laser. A laser light focuses on a fiber and the fiber diameter is measured by processing the scattered laser light. Using the drawing control technique and the precise preform fabrication technique, precision of fiber dimensions to about 0.1 μm has been achieved for today's fibers. This is very important for passive optical components because fiber dimensions determine some of their performance for many components. In most splice methods, the accuracy of fiber core and outer diameter is critical for determining a splice loss. Usually fiber drawing is accompanied by fiber coating (plastic coating) for protection.

The preform is made using several vapor deposition methods. Typical methods are the outer vapor deposition (OVD) method [18,19], the modified chemical vapor deposition (MCVD) method [20], and the vapor-phase axial deposition (VAD) method [21], which are shown in Figure 1.9. In the OVD method [22], submicron-sized

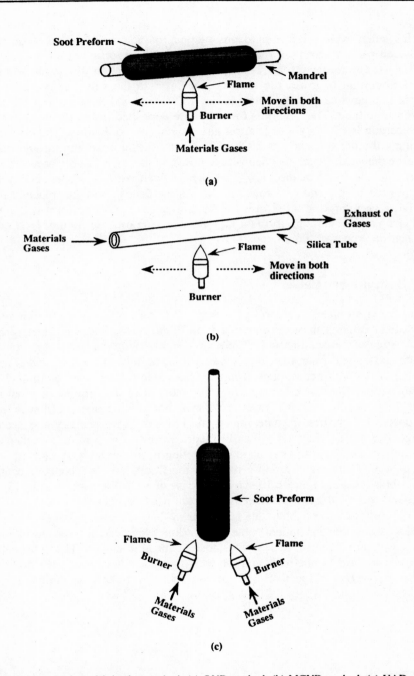

Figure 1.9 Fiber preform fabrication method: (a) OVD method; (b) MCVD method; (c) VAD method.

particles generated by a flame are deposited on a rotating rod (mandrel). The mandrel is removed from a soot preform before the sintering process and a soot preform is dehydrated. The removal of OH (the ion with a combination of oxygen and hydrogen) is important to lower the transmission loss. In the sintering process, a soot preform changes to a glass preform. In the MCVD method [23], a silica tube is used as a substrate tube. The materials for fibers are deposited inside the silica tube, and the deposition is made by using carrier gases such as O_2 or helium. The carrier gases containing the dopants such as $SiCl_4$, $GeCl_4$, and SiF_4 are passing through the tube and are exhausted. After the deposition process, the tube is collapsed to make a preform. In the VAD method [24], submicron-sized particles generated by flames are deposited on the end of a rotating rod. As the deposition takes place at the end of a rod, a very large soot preform can be fabricated. A soot preform is dehydrated and it changes to a glass preform through the sintering process. In the VAD method, these can be processed continuously.

1.2.3 Optical-Fiber Cable

Optical fibers are used in a form of optical cable for optical transmission systems. Optical-fiber cable is shown schematically in Figure 1.10. There are many proposed and investigated cable structures. There are roughly two types of cables used today for a coated optical-fiber structure. One is the assembly of monocoated fibers and the other is that of fiber ribbons. Optical-fiber cable (abbreviated as optical cable) with high-count fibers is composed of units. Contained in a unit are several mono-coated fibers or fiber ribbons. In the case of a monocoated fiber, a fiber is individually coated by plastics. A bare fiber with 125 μm has a primary and secondary coating. The coated diameter typically ranges from 0.7 to 1 mm. The mechanical strength of monocoated fiber in an axial direction is around 8 kg. A fiber ribbon has the structure of aligned fibers in a row and the fibers have a secondary coating to form a ribbon shape. Figure 1.10 shows the case of an 8-fiber ribbon as an example. At present, the 4-, 8-, and 12-fiber ribbons are used practically.

The outlook of optical cables is similar to that of metallic cables, and similar treatment is possible for optical-cable installation. However, the transmission properties of optical cables are superior to those of metallic cables. The outer cable diameter is one-third to one-half, and the weight is one-fifteenth to one-eighth of the diameter and weight, respectively, of metallic cables with the same number of copper pairs.

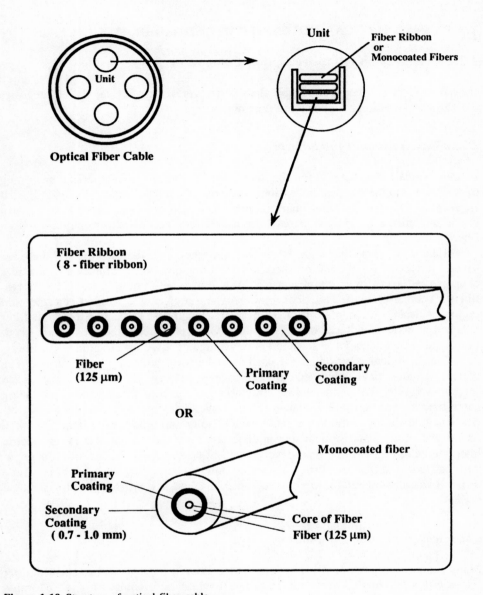

Figure 1.10 Structure of optical-fiber cable.

1.3 PASSIVE OPTICAL COMPONENTS AND THEIR APPLICATION

1.3.1 Classification of Passive Optical Components

Passive optical components can be classified several ways. Here, three categories are shown: structure, function, and port number.

Classification According to Structure

Passive optical components used for fiber transmission can be categorized according to structure into three types: bulk, fiber, and waveguide. The bulk type, as indicated from its name, is bulky, and many traditional optical components such as a prism and a half mirror are used as parts. Lenses are frequently used for coupling with a fiber of this type.

Fiber-type components are fabricated from the fibers themselves, and there are several processes for generating specific fiber functions. For example, a fused-fiber coupler used as a simple coupler or a WDM component is fabricated by fusing two fibers. Another example is a fiber grating, where a grating is formed near the core region. A fiber with miniaturized optical parts, such as a dielectric thin-film filter attached at a fiber endface, belongs to this type. Since fibers themselves are used, fiber-type components have an inherent advantage of low-loss coupling with fibers.

Planar optical waveguides are used to make waveguide-type components. The typical planar optical waveguide is a silica glass waveguide or an ion-exchanged glass waveguide. The light-guiding structures and specific structures for producing some functions are formed in planar optical waveguides by designing the refractive-index profile. Many components, such as a coupler and a passive optical filter, can be integrated in planar optical waveguides. Using a process similar to an electric integrated circuit (IC), this type is very small in size and has the potential for high-volume, low-cost manufacturing. However, low-loss coupling with fibers is a little difficult when compared to fiber-type components, and is very important for practical use.

Classification According to Function

Many functions are required for optical components in order to realize specific fiber transmission systems. Wavelength-selective functions are required for WDM transmission systems. The splitting of optical power is required for forming passive optical network (PON) systems and some optical LAN systems. Some functions can also be realized by both an active and a passive optical component. Although several functions cannot be realized passively, passive optical components play an important role in fiber transmission systems. Several functions required for passive optical components are listed in Table 1.2. Not all the functions required for optical components are listed, just those treated in this book.

Simple connection is the basic function for passive optical components, and it can be categorized into permanent connection and nonpermanent connection (reconnection). Termination terminates the output port of passive optical components without causing reflection light. As the reflection is generated by index mismatching, termination is made by using index matching and light absorption. Attenuation is based on light absorption and is realized by using a light absorber at the wavelength the fiber transmission system used. Generally, attenuation of absorber materials varies with wavelength. When precise attenuation is required, the system designer or user must pay attention to the wavelength. Attenuation is also realized by poor coupling between fibers. The speed of light is finite; therefore, propagation delay takes place when light propagates. Delay of light τ depends both on the refractive index n and on the light propagation length L:

$$\tau = L/v = Ln/c \tag{1.8}$$

where v and c are the light speed in a medium and in a vacuum, respectively. Delay

Table 1.2
Classification of Passive Optical Components According to Function

Function	Explanation of Function	Components
Simple connection	Permanent or difficult to reconnect	Splice
	Easy to reconnect	Optical connector
Termination	No reflection using index matching and attenuation	Optical terminator
Attenuation	Attenuation of optical power	Optical attenuator
Delay	Propagation time delay	Optical delay line
Directional coupler	Coupler with specific coupling depending on propagation directions (Chapter 10)	Optical directional coupler
Split and combine	Optical power splitting and combining	Star coupler
Isolation	No loss for one propagation direction and infinite loss for counterdirection (ideal case)	Optical isolator
Circulation	Coupler with specific coupling depending on propagation directions (Chapter 10)	Optical circulator
Filtering	Selection from lightwaves for propagating or reflection	WDM filter, OFDM filter, band rejection filter, high-pass filter, etc.
Switching	Passing or blocking of propagation light (ON/OFF) or changing the output port of propagating light (routing)	Mechanical switch

is realized by using a fiber or a waveguide with length L. Delay causes the phase shift of light when compared to the original nondelayed light. The directional coupler is very familiar in the field of microwave technology. The optical directional coupler has the same function as the microwave directional coupler. In fiber transmission systems, fiber-type and waveguide-type directional couplers are commonly used. It is also possible to use bulk-type directional couplers, and the directional coupling is realized by using a half mirror in a bulk-type coupler. Generally, bulk-type directional couplers are only used in a laboratory in the case of single-mode fiber transmission systems because of the large coupling loss with a fiber. In the case of multimode fiber transmission systems, bulk-type directional couplers have been used practically. The fused-fiber and waveguide couplers are commonly used for practical single-mode fiber transmission systems. Some systems such as PON require optical power splitting and combining functions, and these functions are realized by a star coupler. Some applications require a wavelength-independent splitting ratio for a star coupler and some require a wavelength-dependent ratio. Both the isolator and the circulator are familiar in the field of microwave technology, and the functions of isolation and circulation in the fiber-optic field are the same as those in the microwave field. Characteristics of laser diodes, which are used for a light source in fiber transmission systems, are affected by the injection light. The linewidth of the laser light spectrum varies according to the reflected light from the passive optical components. To avoid this, some systems use isolators. The circulator is not so commonly used in fiber transmission systems; however, several applications using the circulator, including an erbium-doped fiber amplifier (EDFA), are considered. Optical filtering is the critical function for realizing WDM and OFDM systems. It separates the specific wavelength light among multiplexed lights with many wavelengths. It is also used as the combiner of lights with the specific wavelength. Optical filtering is considered to be a promising function for optical signal processing. Optical switching is the realization of switching in the optical domain, and this function is generally realized by active components such as semiconductor devices. However, many mechanical switches have been realized based on the splice and connector technologies. The major difference between these mechanical switches and simple connection components (splice and connector) is the driving mechanism. Therefore, a mechanical switch is included as a passive component in this book.

Since the simple connection is the basic function for passive optical components, the classifications of splice and connector are shown in detail in Tables 1.3 and 1.4. Splices are categorized into three major types: fusion, adhesive, and mechanical splices. Fusion splices use heat for fusing two mated fibers, and fusion splice methods can be subdivided according to the heat source. The heat sources investigated so far are an electric discharge, a gas laser, a flame, and an electric heater. Among them, the electric discharge is commonly used. Neither adhesive nor mechanical splices use heat, and the spliced state is maintained by adhesion of chemical bond or mechanical force, respectively. In Table 1.3, the applicability of these

Table 1.3
Classification of Splicing Methods

	Fiber Type		
Splice Method	*Silica Fiber*	*Fluoride Fiber*	*Plastic Fiber*
Permanent			
Fusion splice			
Electric discharge	Possible	Possible	Difficult
Gas laser (CO_2)	Possible	Possible	Difficult
Flame	Possible	Possible	Difficult
Electric Heater	Impossible	Possible	Possible
Adhesive splice	Possible	Possible	Possible
Nonpermanent (possible to reconnect)			
Mechanical splice	Possible	Possible	Possible

Table 1.4
Classification of Connectors According to
Simultaneous Connection Number

Optical Connector	
Single fiber	Single-fiber connector
Multifiber	Hybrid of single-fiber connector (2–20)
	Array-fiber connector (2–20)
	2D fiber connector (20–200)

splice methods to several fiber types is also shown. Fusion splicing methods using an electric heater cannot apply to silica fibers because of the lack of heating temperature. On the other hand, other fusion splicing methods are difficult to apply to plastic fibers because of the high heating temperature.

Optical connectors can be classified according to the simultaneous connection number of fibers (Table 1.4). Ordinarily, single-fiber connectors are used. Multifiber connectors such as an array-fiber connector are frequently used for fiber-ribbon cables, and they realize the compact and efficient connection. Two-dimensional (2D) fiber connectors are fibers in a connector ferrule that are two-dimensionally positioned like a lattice. In this sense, array-fiber connectors are one-dimensional (1D) connectors, where fibers are positioned in a row. In Table 1.4, the number in parentheses indicates the typical simultaneous fiber connection number. Optical connectors can be also classified according to the materials used for connectors. The materials are metal, glass, silicon, plastics, ceramics, and their compounds.

Classification According to Port Number

Passive optical components can be classified according to input and output port number (Table 1.5). Star couplers and WDM filters use many ports for the splitting or separation of lightwaves. In some applications where the reflection of light causes serious degradation of system performance, terminators may be used for the unused ports.

Table 1.5
Classification of Passive Optical Components
According to Port Number

Port Number	Examples of Passive Optical Components
1-port	Termination
2-port	Splice, connector, attenuator, delay line, filter
3-port	WDM filter, OFDM filter
4-port	WDM filter, OFDM filter, directional coupler
n-port	WDM filter, OFDM filter, star coupler

1.3.2 Mathematical Representation of Passive Optical Components

Passive optical components are represented mathematically by using several matrix formats. Among them, a scattering matrix is used in this book. The scattering matrix is commonly used in the field of microwave technology [25], and it is also used with the same definition in fiber optics. General passive optical components with n-ports are shown in Figure 1.11. The two complex amplitudes a_i and b_i for port i are used. The notations a_i and b_i represent the complex amplitudes for the input and the reflected lightwaves, respectively. For the n-port case, the relation between the input and the reflected lightwaves is represented as

$$\begin{bmatrix} b_1 \\ \cdot \\ \cdot \\ \cdot \\ b_n \end{bmatrix} = \begin{bmatrix} s_{11} & \cdot & \cdot & s_{1n} \\ \cdot & \cdot & \cdot & \cdot \\ \cdot & \cdot & \cdot & \cdot \\ s_{n1} & \cdot & \cdot & s_{nn} \end{bmatrix} \begin{bmatrix} a_1 \\ \cdot \\ \cdot \\ \cdot \\ a_n \end{bmatrix} \tag{1.9}$$

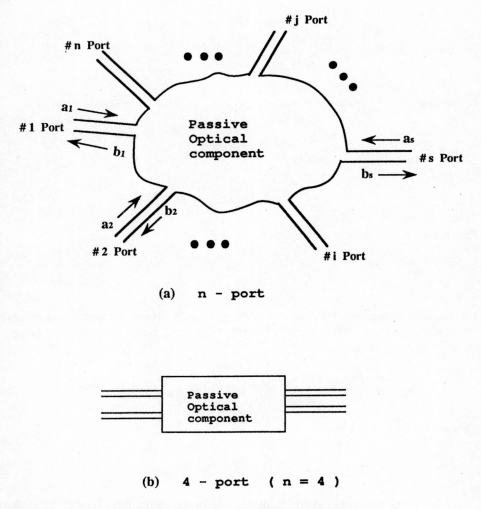

Figure 1.11 *n*-port passive optical component: (a) *n*-port; (b) 4-port (*n* = 4).

where $\mathbf{b} = [b_1, b_2, \ldots, b_n]$ and $\mathbf{a} = [a_1, a_2, \ldots, a_n]$ are the vectors, and $[S]$ is the scattering matrix. This equation is also expressed using vectors and the scattering matrix as

$$\mathbf{b} = [S]\mathbf{a} \tag{1.10}$$

In the case of 4-port components, $[S]$ is

$$[S] = \begin{bmatrix} s_{11} & s_{12} & s_{13} & s_{14} \\ s_{21} & s_{22} & s_{23} & s_{24} \\ s_{31} & s_{32} & s_{33} & s_{34} \\ s_{41} & s_{42} & s_{43} & s_{44} \end{bmatrix} \qquad (1.11)$$

In the case of the components with reciprocity, the following equation holds:

$$S_{ij} = S_{ji} \qquad (1.12)$$

The input into the component is

$$P_{in} = \sum_{j=1}^{n} |a_j|^2 = \sum_{j=1}^{n} a_j a_j^* \qquad (1.13)$$

The output power from the component is

$$P_{out} = \sum_{j=1}^{n} |b_j|^2 = \sum_{j=1}^{n} b_j b_j^* \qquad (1.14)$$

The notation z^* represents the complex conjugate of z, where z is a complex number or variable. When the component is lossless, $P_{in} = P_{out}$ holds from the energy conservation law. That is,

$$P_{out} = \sum_{j=1}^{n} \left(\sum_{i=1}^{n} s_{ji} a_i \right) \left(\sum_{k=1}^{n} s_{jk}^* a_k^* \right) = \sum_{i=1}^{n} \sum_{k=1}^{n} \left(\sum_{j=1}^{n} s_{ji} s_{jk}^* \right) a_i a_k^* = P_{in} \qquad (1.15)$$

This results in

$$\sum_{j=1}^{n} s_{ji} s_{jk}^* = \delta_{ik} \qquad (1.16)$$

where δ_{ik} is the Kronecker delta, with $\delta_{ik} = 1$ in the case of $i = j$, and $\delta_{ik} = 0$ in the case of $i \neq j$. Equation (1.16) always holds for lossless components, and the scattering matrix satisfying this equation is called a unitary matrix. The scattering matrix is very useful because the input and reflected lightwave can be considered in one matrix.

1.3.3 Application for Fiber Transmission

Passive optical components are used in many fiber transmission systems. Simple connections such as splices and connectors are inevitable for all fiber transmission systems ranging from trunk (long-distance) systems to LAN (localized and short-distance) systems. An optical attenuator is used, for example, when the transmission distance is too short and the laser light is too strong for the receiver. Isolators are sometimes used in front of a laser diode for high-speed transmission systems where the reflected light degrades the system performance. Optical couplers including a star coupler are used in some systems, and typical usages are shown in Figure 1.12. Figure 1.12(a) shows the bidirectional transmission system using one fiber. Optical couplers are used in transceivers for input and output of laser light. Other cases in the figure use couplers as branching and combining components.

Optical filters such as WDM and OFDM filters are used in WDM and OFDM systems as key components (Fig. 1.13(a–c)). In Figure 1.13(c), WDM filters are used as both the combiner and the separator of lightwaves. Optical band rejection filters are used in a testing system (Fig. 1.13(d)). In Figure 1.13(d), the light with λ_2 is used for testing fiber lines and the light with λ_1 for signal transmission. Important future applications of optical switches are optical exchanges and optical cross connects. Mechanical switches are difficult to use in these systems because of their low switching speed. Possible applications of mechanical switches for fiber transmission systems are shown in Figure 1.14. These are the switches for the protection of transceivers and for fiber routing. In this section, applications of passive optical components are explained only for fiber transmission systems. Although not explained in this book, many passive optical components are used in optical measuring systems.

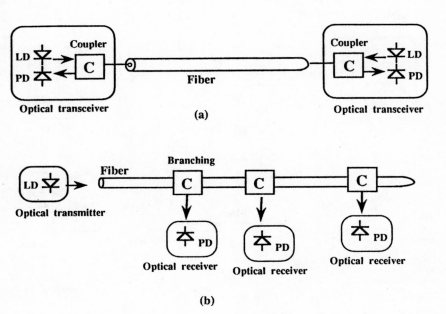

Figure 1.12 Application of optical couplers in fiber transmission.

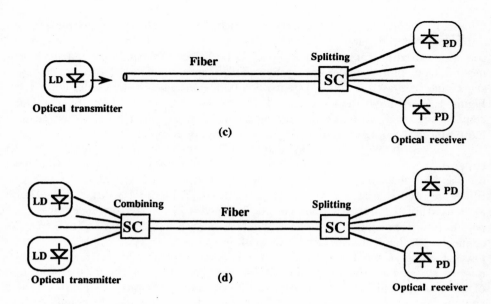

Figure 1.12 (Continued)

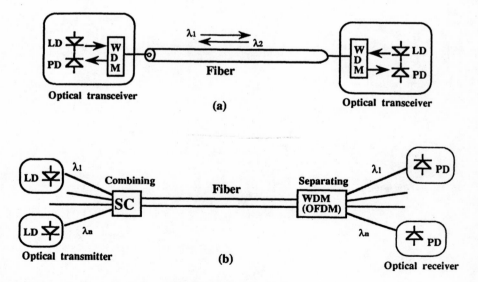

Figure 1.13 Application of optical filters in fiber transmission.

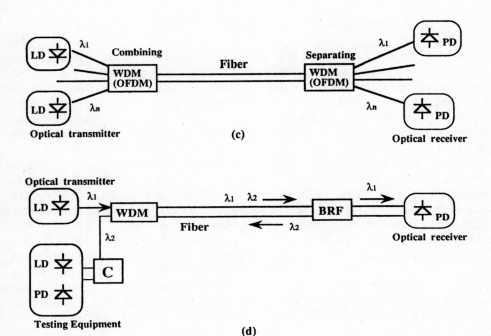

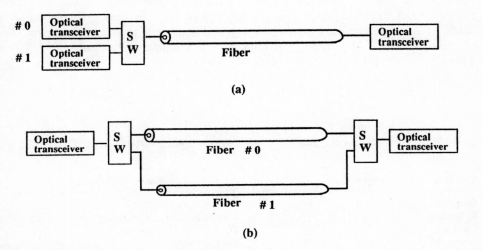

Figure 1.14 Application of optical switch in fiber transmission.

1.4 REQUIREMENTS FOR PASSIVE OPTICAL COMPONENTS

Passive optical components are used in several fiber transmission systems, as explained in the previous section. Various characteristics, depending on the systems, are required for the components. Some of these requirements may be peculiar to specific components, such as the isolation required for an isolator. Several requirements may be common, and these are considered in this section. The relationship between the characteristics of passive components and system performance is shown in Table 1.6.

Insertion losses of passive components increase the transmission loss and influence the available transmission distance. Insertion loss is required to be low for all passive components except for an attenuator. Reflection from passive components causes system instability and signal distortion in some systems. For example, the reflected light may inject a laser diode and cause unstable lasing. The injected light may cause the nonlinearity in optical modulation and may result in signal distortions in optical analog transmission systems. Reflection also deteriorates the system performance in such a bidirectional system, as indicated in Figure 1.15(a). In this system, the same wavelength is used for up and down streams. Crosstalk generally degrades the error rate of the systems, and an example of this for a WDM system is shown in Figure 1.15(b). The number of passive components used in a system is usually much larger than the number of active components. Therefore, the cost of passive components has a strong influence on the system cost. The environment, such as temperature change, mechanical vibrations, and humidity, has considerable influence on the performance of passive components. Typical influenced characteristics are the insertion loss and lifetime of passive components. To ensure system reliability and stability, highly reliable components must be used. When used in an outside plant, passive components are exposed to a very severe environment. For example, insertion loss stability is required for a temperature change of $-30°$ to $+60°C$.

Table 1.6
Relationship Between Components and System Performance

Characteristics of Components	*System Performance*
Insertion loss	Available transmission distance
Reflection	System instability, distortion of signals (analog)
Crosstalk	Degradation of error rate
Cost	System cost
Reliability against environmental change	System reliability, system instability

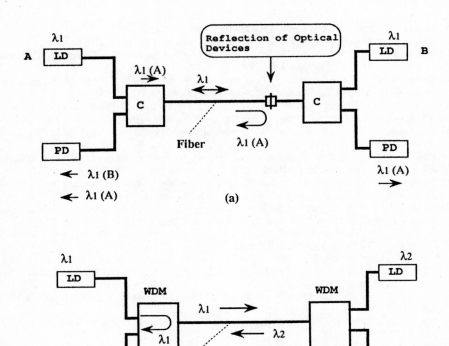

Figure 1.15 (a) Optical reflection and (b) crosstalk.

REFERENCES

[1] Maiman, T. H., "Stimulated Optical Radiation in Ruby Masers," Nature, Vol. 187, 1960, p. 493.

[2] Kao, K. C., and G. A. Hockham, "Dielectric-Fiber Surface Waveguides for Optical Frequencies," Proc. IEEE, Vol. 113, 1966, p. 1151.

[3] Hayashi, I., M. B. Panish, P. W. Foy, and A. Sumski, "Function Lasers Which Operate Continuously at Room Temperature," Appl. Phys. Lett., Vol. 17, 1970, p. 109.

[4] Kapron, F. P., D. B. Keck, and R. D. Maurer, "Radiation Losses in Glass Optical Waveguides," Appl. Phys. Lett., Vol. 17, 1970, p. 423.

[5] Miya, T., Y. Terunuma, T. Hosaka, and T. Miyashita, "Ultimate Low-Loss Single Mode Fibers at 1.55 μm," Electron. Lett., Vol. 15, 1979, p. 106.

[6] Kashima, N., Optical Transmission for the Subscriber Loop," Norwood, MA: Artech House, 1993.

[7] Martin, J., Local Area Networks, Prentice-Hall, 1989.

[8] Gloge, D., "Weakly Guiding Fibers," Appl. Opt., Vol. 10, 1971, p. 2252.

[9] Petermann, K., "Constraints for Fundamental-Mode Spot Size for Broadband Dispersion-Compensated Single-Mode Fibers," Electron. Lett., Vol. 19, 1983, p. 712.

[10] Ramaswamy, V., I. P. Kaminow, P. Kaiser, and W. G. French, "Single Polarization Optical Fibers: Exposed Cladding Technique," Appl. Phys. Lett., Vol. 33, 1978, p. 814.

[11] Dyott, R. B., J. R. Cozens, and D. G. Morris, "Preservation of Polarization in Optical-Fiber Waveguides With Elliptical Cores," Electron. Lett., Vol. 15, 1979, p. 380.

[12] Katuyama, T., H. Mastumura, and T. Suganuma, "Low-Loss Single Polarization Fibers," Electron. Lett., Vol. 17, 1981, p. 473.

[13] Hosaka, T., K. Okamoto, T. Miya, Y. Sasaki, and T. Edahiro, "Low-Loss Single Polarization Fibers With Asymmetric Strain Birefringence," Electron. Lett., Vol. 17, 1981, p. 530.

[14] Okoshi, T., K. Oyamada, M. Nishimura, and H. Yokota, "Side-Tunnel Fiber: An Approach to Polarization-Maintaining Optical Waveguiding Scheme," Electron. Lett., Vol. 18, 1982, p. 824.

[15] Birch, R. D., D. N. Payne, and M. P. Varnham, "Fabrication of Polarization-Maintaining Fibers Using Gas-Phase Etching," Electron. Lett., Vol. 18, 1982, p. 1036.

[16] Stolen, R. H., W. Pleibel, and J. R. Simpson, "High-Birefringence Optical Fibers by Preform Deformation," IEEE J. Lightwave Technol., Vol. LT-2, 1984, p. 639.

[17] Yariv, A., Quantum Electron, John Wiley & Sons, 1975.

[18] Keck, D. B., P. C. Schultz, and F. Zimar, "Method of Forming Optical Waveguide Fibers," U.S. Pat. 3,737,292, 1973.

[19] Maurer, R. D., " Method of Forming an Economic Optical Waveguide Fiber," U.S. Pat. 3,737,293, 1973.

[20] MacChesney, J. B., and P. B. O'Connor, "Optical Fiber Fabrication and Resultant Product," U.S. Pat. 4,217,027, 1980.

[21] Izawa, T., T. Miyashita, and F. Hanawa, "Continuous Optical Fiber Preform Fabrication Method," U.S. Pat. 4,062,665, 1977.

[22] Schultz, P. C., "Fabrication of Optical Waveguides by the Outside Vapor Deposition Process," Proc. IEEE, Vol. 68, 1980, p. 1187.

[23] Nagel, S. R., J. B. MacChesney, and K. L. Walker, "An Overview of the Modified Chemical Vapor Deposition (MCVD) Process and Performance," IEEE J. Quantum Electron., Vol. QE-18, 1982, p. 459.

[24] Izawa, T., and N. Inagaki, "Materials and Processes for Fiber Preform Fabrication Vapor-Phase-Axial Deposition," Proc. IEEE, Vol. 68, 1980, p. 1184.

[25] Collin, R. E., Foundations of Microwave Engineering, McGraw-Hill, 1966.

Chapter 2
Characteristics of Optical Fibers

This chapter deals with the characteristics of silica fibers as the basis of passive optical components. Transmission characteristics of both single-mode and multimode fibers are briefly explained with electromagnetic theory based on Maxwell's equations. Many modes exist in multimode fibers, and mode conversions take place in the fibers themselves and in passive optical components attached to multimode fibers. Mode conversion makes transmission characteristics complex, and it will be explained in Chapter 3 using an example of mode conversion caused by a splice or a connector. Although silica fiber is a glass, it is strong enough for practical use, but becomes brittle when it has a damaged or scratched surface. These transmission and mechanical characteristics are explained here.

2.1 TRANSMISSION CHARACTERISTICS OF FIBERS

2.1.1 Basic Equations for Fiber Transmission Characteristics

The basic equations for dielectric waveguides like fibers are Maxwell's equations. The following three assumptions are made using Maxwell's equations:

1. Permeability is equal to the vacuum value because of the light frequency range. The refractive index is n. That is,

$$\mu = \mu_0 \quad \text{and} \quad \varepsilon = \varepsilon_0 n^2 \tag{2.1}$$

where μ_0 and ε_0 are the permeability and permittivity of a vacuum, respectively.
2. There is no source.
3. The time t and the z-coordinate (propagation direction) dependence is

$$\exp(j\omega t - \gamma z) \tag{2.2}$$

where ω and γ are the angular frequency and propagation constant, respectively. Maxwell's equations are

$$\text{div}(\varepsilon_0 n^2 \mathbf{E}) = 0$$

$$\text{div}(\mu_0 \mathbf{H}) = 0 \tag{2.3}$$

$$\text{rot } \mathbf{E} = -j\omega\mu_0 \mathbf{H}$$

$$\text{rot } \mathbf{H} = j\omega\varepsilon_0 n^2 \mathbf{E}$$

The notation rot is the same as curl. $\mathbf{E}$ and $\mathbf{H}$ in (2.3) can be divided into the cross-sectional components $\mathbf{E}_t$ and $\mathbf{H}_t$ and the axial components E_z and H_z. The equations for components $\mathbf{E}_t$ and $\mathbf{H}_t$ are

$$\nabla^2 \mathbf{E}_t + (\omega^2 n^2 \varepsilon_0 \mu_0 - \beta^2)\mathbf{E}_t + \nabla\left[\frac{\nabla n^2}{n^2} \cdot \mathbf{E}_t\right] = 0 \tag{2.4}$$

$$\nabla^2 \mathbf{H}_t + (\omega^2 n^2 \varepsilon_0 \mu_0 - \beta^2)\mathbf{H}_t + \left(\frac{\nabla n^2}{n^2}\right) \times (\nabla \times \mathbf{H}_t) = 0 \tag{2.5}$$

Here, $\gamma = \alpha + j\beta$ (α = loss constant, β = phase constant) and $\alpha = 0$ is assumed. Equations (2.4) and (2.5) are called vector wave equations [1]. The axial components E_z and H_z are related to the solutions $\mathbf{E}_t$ and $\mathbf{H}_t$ of (2.4) and (2.5). When ∇n^2 is small, (2.4) and (2.5) can be treated as scalar wave equations, such as

$$\nabla^2 \mathbf{E}_t + (\omega^2 n^2 \varepsilon_0 \mu_0 - \beta^2)\mathbf{E}_t = 0 \tag{2.6}$$

$$\nabla^2 \mathbf{H}_t + (\omega^2 n^2 \varepsilon_0 \mu_0 - \beta^2)\mathbf{H}_t = 0 \tag{2.7}$$

The solutions of wave equations are called modes. Among the modes, the following orthogonal relation is known to hold:

$$\int_S \mathbf{E}_t^i \times \mathbf{H}_t^{j*} \, dS = 0 \qquad (i \neq j) \tag{2.8}$$

where integration is made for the fiber cross-sectional plane.

The wave equation with boundary conditions (at the core-cladding boundary) derives the eigenvalue equation, and its solution determines the β value. With β, the group delay τ for a unit fiber length is expressed as

$$\tau = \frac{1}{v_g} = \frac{d\beta}{d\omega} \tag{2.9}$$

where v_g is the group velocity. The bandwidth of a fiber is determined by the group delay τ and there are three factors influencing the bandwidth:

1. Different group velocities for modes (modal dispersion);
2. Wavelength-dependent refractive index of fiber materials (material dispersion);
3. Wavelength dependency of the group velocities caused by the fiber structure (waveguide dispersion).

In multimode fibers, the modal dispersion is dominant and the remaining dispersions are negligible. Since there exists only one mode for single-mode fibers in the case of ordinary usage (there are no multimode dispersions), both material and waveguide dispersions are important.

2.1.2 Optical Loss of a Fiber

Along with bandwidth, optical loss is another important transmission characteristic. Loss $\alpha(\lambda)$ in a straight fiber state is expressed by (2.10) as a function of wavelength λ.

$$\alpha(\lambda) = \frac{C_1}{\lambda^4} + C_2 + A(\lambda) \tag{2.10}$$

where C_1 and C_2 are constants and $A(\lambda)$ is a function of λ. The three terms correspond to the following three major origins:

1. Rayleigh scattering loss due to microfluctuation of refractive index (the first term);
2. Loss due to imperfections in fiber structure (the second term);
3. Loss due to impurities and intrinsic ultraviolet (UV) and infrared (IR) absorption (the third term).

Microfluctuation of the refractive index, which is smaller than the used light wavelength, causes Rayleigh scattering. The origin of microfluctuation is the fluctuation of density or composition in a glass. This scattering is known to be proportional to λ^{-4}, and the fiber loss decreases greatly for longer wavelengths when this scattering is dominant. Constant C_1 equals 0.6 to 0.8 μm^4 dB/km for today's single-mode fibers. Rayleigh-scattered light scatters in all directions, and the backward light from Rayleigh scattering is used for the optical time-domain reflectometer (OTDR). In the usual OTDR, a short optical light pulse is injected into fibers. With OTDR, fiber and component loss such as a splice loss can be measured as a function of fiber length. It is possible to measure the two component losses separately when the injected optical light pulse is sufficiently short. The second term is the loss due to imperfection in fiber structure, such as microbending and core-cladding imperfection. This loss mechanism is wavelength-independent. Constant C_2 is very small for today's fibers because of improvements in fiber manufacturing technology. $A(\lambda)$ represents the loss of impurities and intrinsic UV and IR absorption. OH and metal ions are the major impurities in silica fibers. Metal ions are not a serious problem for today's fibers, again because of the improvements in fiber manufacturing technology. Loss peaks due to OH exist at wavelengths of 0.95, 1.13, 1.24, 1.39, 1.90, and 2.22 μm in the 0.9- to 2.5-μm wavelength band. OH has also been eliminated in good fibers, but a small amount still exists. OH affects the long-term reliability on fiber loss. It is known that OH is made from diffused hydrogen gas and constituent oxygen atoms after optical-cable installation. Loss values due to this mechanism depend on dopant types and optical-cable materials. Today's optical transmission systems are constructed keeping this mechanism in mind, so they work well. UV absorption originates from electronic band gap transitions, and the loss due to this absorption decreases at longer wavelengths. IR absorption, which originates from multiphonon absorption (molecular vibration), increases at longer wavelengths. The wavelength, where loss is minimum, depends on the first and third terms in (2.10). Typical transmission losses as a function of wavelength are shown in Figure 2.1. The two curves represent OH existing and OH free single-mode fibers, respectively. The reported minimum transmission loss is 0.15 dB/km at 1.55 μm [2]. Equation (2.10) describes the fiber loss without a bend. When fibers are bent, additional loss must be considered. However, the bending loss is negligibly small for the bending radius $R > 10$ cm for usual fibers.

The structural design of fiber is also important for obtaining low-loss fiber and optical cables. Light injected from a light source such as a laser diode is guided by a core-cladding structure. Most of the guided light energy exists in the core, but a small part of the light energy is in the cladding [3]. Therefore, material loss of the cladding as well as that of the core must be low to obtain low-loss fibers. It is also important that the cladding diameter be suitably designed [4]. Because of the fact that a small part of light exists in the cladding, several coupling components can be devised, such as a light-tapping component from a fiber and a fused-fiber coupler. In a fused-fiber coupler, light coupling between cores occurs because of the small distance between cores. This is analogous to a microwave directional coupler using strip lines, where electromagnetic fields leak from one strip line to another.

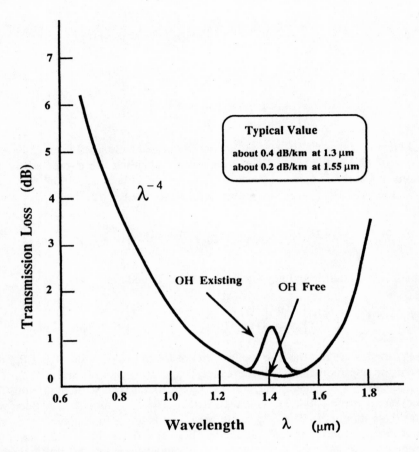

Figure 2.1 Spectral loss of single-mode fiber.

2.2 TRANSMISSION CHARACTERISTICS OF SINGLE-MODE FIBERS

2.2.1 Mode Field of Single-Mode Fiber

Scalar wave equations (2.6) and (2.7) for $\mathbf{E}_t$ and $\mathbf{H}_t$ can be rewritten with the axial components (longitudinal components) E_z or H_z using the fact that $\mathbf{E}_t$ and $\mathbf{H}_t$ are related to E_z and H_z. The rewritten result of the scalar wave equation is expressed using a cylindrical coordinate (r, ϕ, z) [5,6]:

$$\frac{\partial^2 \psi}{\partial r^2} + \frac{1}{r}\frac{\partial \psi}{\partial r} + \frac{1}{r^2}\frac{\partial^2 \psi}{\partial \phi^2} + (n^2 k^2 - \beta^2)\psi = 0 \qquad (2.11)$$

where

$$k^2 = \omega^2 \varepsilon_0 \mu_0 = \left(\frac{2\pi}{\lambda}\right)^2 \qquad (2.12)$$

In (2.11), ψ represents E_z or H_z, and other fields can be obtained from E_z or H_z. The solutions of (2.11) for an SI fiber with infinite cladding are expressed by Bessel functions in a core region and by Hankel functions in a cladding region, respectively. That is, the field of a core region ($r \leq a$) for an $LP_{\nu\mu}$ mode

$$E_z = A \left\{ J_{\nu+1}(\kappa r) \begin{bmatrix} \sin(\nu+1)\Phi \\ -\cos(\nu+1)\Phi \end{bmatrix} + J_{\nu-1}(\kappa r) \begin{bmatrix} \sin(\nu-1)\Phi \\ -\cos(\nu-1)\Phi \end{bmatrix} \right\} \qquad (2.13)$$

and that for a cladding region ($r \geq a$) for an $LP_{\nu\mu}$ mode

$$E_z = B \left\{ H_{\nu+1}^{(1)}(j\gamma r) \begin{bmatrix} \sin(\nu+1)\Phi \\ -\cos(\nu+1)\Phi \end{bmatrix} + H_{\nu-1}^{(1)}(j\gamma r) \begin{bmatrix} \sin(\nu-1)\Phi \\ -\cos(\nu-1)\Phi \end{bmatrix} \right\} \qquad (2.14)$$

The $LP_{\nu\mu}$ modes are not the true modes, except for the LP_{01} mode. The $LP_{\nu\mu}$ modes represent the approximations for the exact $TE_{0\mu}$ (transverse electric), $TM_{0\mu}$ (transverse magnetic), $HE_{\nu\mu}$, and $EH_{\nu\mu}$ modes (hybrid modes) in weakly guiding fibers, and they are a combination of the exact modes. The $LP_{\nu\mu}$ modes have been widely used for their simplicity. The LP_{01} mode corresponds to the fundamental HE_{11} mode, and it is the exact mode in a single-mode fiber. The fields that are the solutions in both regions must satisfy the boundary conditions at the core-cladding boundary ($r = a$). This derives the following eigenvalue equation for a weakly guiding SI fiber.

$$\kappa \frac{J_{\nu+1}(\kappa a)}{J_\nu(\kappa a)} = j\gamma \frac{H_{\nu+1}^{(1)}(j\gamma a)}{H_\nu^{(1)}(j\gamma a)} \qquad (2.15)$$

where

$$\kappa = (n_1^2 k^2 - \beta_\nu^2)^{1/2} \qquad (2.16)$$

$$\gamma = (\beta_\nu^2 - n_2^2 k^2)^{1/2} \qquad (2.17)$$

Here, $n_1 = n_0$ (constant) because of an SI fiber. The normalized frequency V is expressed by κ and γ. That is,

$$V^2 = (\kappa a)^2 + (\gamma a)^2 = (n_1^2 - n_2^2)k^2 a^2 \qquad (2.18)$$

which is equal to (1.6). The solution of the eigenvalue equation determines β and cutoff frequency V_C. Modes reach cutoff when $\gamma = 0$. For $\gamma \to 0$, (2.15) is rewritten as

$$V_C \frac{J_{\nu+1}(V_C)}{J_\nu(V_C)} = jx \frac{H_{\nu+1}^{(1)}(jx)}{H_\nu^{(1)}(jx)} \qquad (2.19)$$

where x is a very small value ($x \to 0$) and $V_C = (\kappa a)_C$ for $\gamma = 0$ is used.

For a small z value, the following approximation for the Hankel function holds [7]:

$$H_0^{(1)}(z) = (2j/\pi) \log(z) \qquad (2.20)$$

$$H_\nu^{(1)}(z) = -(j(\nu - 1)!/\pi)(z/2)^{-\nu} \qquad \text{(for } \nu \neq 0) \qquad (2.21)$$

Using these equations, (2.19) for $x \to 0$ can be evaluated. The right-hand side of (2.19) tends to be zero for $x \to 0$ when $\nu = 0$. Therefore, the cutoff condition for $\nu = 0$ is obtained:

$$J_1(V_C) = 0 \qquad \text{(for } \nu = 0 \text{ mode: LP}_{0\mu} \text{ mode)} \qquad (2.22)$$

When $\nu \geq 1$, the right-hand side of (2.19) does not tend to zero for $x \to 0$. Therefore, (2.19) is rewritten by using the following recurrence relation for Bessel and Hankel functions [7]:

$$J_{\nu-1}(z) + J_{\nu+1}(z) = \frac{2\nu}{z} J_\nu(z) \qquad (2.23)$$

$$H_{\nu-1}^{(1)}(z) + H_{\nu+1}^{(1)}(z) = \frac{2\nu}{z} H_\nu^{(1)}(z) \qquad (2.24)$$

and the result is

$$V_C \frac{J_{\nu-1}(V_C)}{J_\nu(V_C)} = jx \frac{H_{\nu-1}^{(1)}(jx)}{H_\nu^{(1)}(jx)} \qquad (2.25)$$

The right-hand side of (2.25) tends to zero for $x \to 0$ when $\nu \geq 1$, and the cutoff condition is

$$J_{\nu-1}(V_C) = 0 \qquad \text{(for } \nu = 1, 2, \ldots \text{ mode: LP}_{\nu\mu} \text{ mode)} \qquad (2.26)$$

The cutoff values are obtained from the Bessel function zeros. These are listed as [7]:

$LP_{0\mu}$: 0, 3.832, 7.016, ...

$LP_{1\mu}$: 2.405, 5.520, 8.654, ...

$LP_{2\mu}$: 0, 3.832, 7.016, ...

$LP_{3\mu}$: 0, 6.380, 9.761, ...

The μ is labeled according to the order number from the lowest zero value. From this list, $V_C = 0$ for LP_{01} and $LP_{\nu1}$ ($\nu \geq 2$). Therefore, LP_{01} and $LP_{\nu1}$ ($\nu \geq 2$) seem to be the lowest or fundamental modes. This is wrong because $V_C = 0$ is not the solution of the eigenvalue equation for $LP_{\nu1}$ ($\nu \geq 2$). When $x \to 0$ and $V_C \to 0$, both sides of (2.19) tend to be zero for $\nu = 0$. On the other hand, the right-hand side of (2.25) tends to be zero for $x \to 0$ and $V_C \to 0$, while the left-hand side of (2.25) tends to be 2ν (in the case of $\nu \geq 2$). This indicates that $V_C = 0$ is not the solution for the $LP_{\nu1}$ ($\nu \geq 2$) mode. The mode that satisfies the single-mode propagation condition is the LP_{01} mode, and this mode has no cutoff. The single-mode propagation condition for an SI fiber is violated when $V \geq 2.405$.

The fundamental mode of single-mode fibers is the LP_{01} (HE_{11}) mode. Strictly speaking, the mode field of this fundamental mode is not a Gaussian profile. As a convenience, it is approximately expressed by the following Gaussian equation [8]:

$$E_y = A \exp\left(-\frac{r^2}{\omega^2}\right) \tag{2.27}$$

where A is a constant and ω is the mode field radius, also called the spot size. The mode field radius ω is expressed as

$$\frac{\omega}{a} = 0.65 + \frac{1.619}{V^{3/2}} + \frac{2.879}{V^6} \qquad \text{(SI)} \tag{2.28}$$

$$\frac{\omega}{a} = \sqrt{\frac{2}{V} + \frac{0.23}{V^{3/2}} + \frac{18.01}{V^6}} \qquad \text{(GI, } \alpha = 2\text{)} \tag{2.29}$$

By using the Gaussian field approximation, the ratio of power in a cladding to the total power is obtained simply as

$$\frac{P_{\text{clad}}}{P_{\text{total}}} = \frac{\int_a^\infty |E_y|^2 \, r \, dr}{\int_0^\infty |E_y|^2 \, r \, dr} = \exp\left(-\frac{2a^2}{\omega^2}\right) \tag{2.30}$$

In the case of $V = 2.4$, $\omega/a = 1.101$ and $P_{\text{clad}}/P_{\text{total}} = 0.19$ for an SI fiber. This result indicates that about 20% of total power exists in the cladding region for $V = 2.4$. The mode field radius ω is very useful for evaluating a coupling loss such as a splice loss [9]. In (2.28) and (2.29), the mode field radius ω is expressed by using the core radius a and the refractive-index difference Δ (normalized frequency V is defined by using Δ). For a fiber with the general index profile, it is difficult to measure a and Δ. Therefore, it is difficult in practice to calculate ω by using a, and Δ (or V) for such a fiber. Practically, ω is obtained using several procedures: by using the near-field pattern (NFP: $\phi(r)$) or the far-field pattern (FFP: $\psi(r)$) of a fiber mode field. NFP and FFP are shown in Figure 2.2. The following equations are proposed for calculating ω with the measured field patterns [10,11].

$$\omega^2 = \frac{2 \int_0^\infty \phi^2(r) r^3 \, dr}{\int_0^\infty \phi^2(r) r \, dr} \tag{2.31}$$

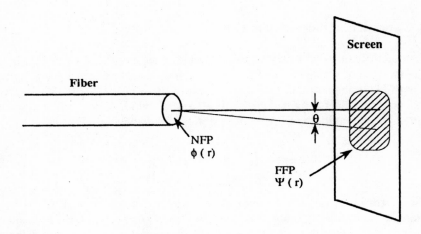

Figure 2.2 Near- and far-field patterns of a fiber.

$$\omega^2 = \frac{2 \int_0^\infty \psi^2 (p)p \, dp}{\int_0^\infty \psi^2 (p)p^3 \, dp} \tag{2.32}$$

where, $p = k \sin \theta$, and θ is the angle shown in Figure 2.2. The FFP of an electric antenna in the Fraunhofer region, which is defined as the region located at a long distance from the antenna, is related to the current distribution (NFP) on the antenna surface by the Fourier transform. The relationship between FFP and NFP is the same in the case of a fiber. The Fourier transform is replaced by the Hankel transform due to the circular cross section of a fiber. In [12], (2.32) is shown to be equal to Petermann's original equation (2.33) by using the Hankel transform:

$$\omega^2 = \frac{2 \int_0^\infty \phi^2 (r)r \, dr}{\int_0^\infty \left(\frac{d\phi}{dr}\right)^2 r \, dr} \tag{2.33}$$

2ω is usually called the mode field diameter (MFD). These two MFD definitions of ω are equal in the case of the Gaussian field, but are different for the general field profile. According to [13], the definition of (2.32) is suitable for splice loss evaluation.

2.2.2 Dispersion of Single-Mode Fiber

Single-mode fibers have a very large bandwidth because of the single mode. However, dispersion of single-mode fibers is still important when considering high-speed and long-fiber transmission. The group delay τ for a unit fiber length is given by (2.9). The delay T of a light pulse with a spectral width $\delta\lambda$ for a unit fiber length is written by expanding near the central optical frequency ω_0. That is,

$$T = \left[\frac{1}{v_g}\right]_{\lambda=\lambda_0} + \left[\frac{d}{d\lambda}\left(\frac{1}{v_g}\right)\right]_{\lambda=\lambda_0} \delta\lambda \tag{2.34}$$

The first term $[1/v_g]_{\lambda=\lambda_0}$ corresponds to the modal dispersion, and this is constant for a single-mode fiber in a single-mode condition. The second term corresponds to the chromatic dispersion, which is composed of material and waveguide dispersions. Here, the chromatic dispersion is considered for an SI single-mode fiber ($n_1 = n_0$ (constant)). Before the calculation, the following parameter b is usually used:

$$b = \frac{(\gamma a)^2}{V^2} = \frac{(\beta/n_2 k)^2 - 1}{(n_1/n_2)^2 - 1} \tag{2.35}$$

The parameter b ranges from 0 to 1. It takes 0 at cutoff ($\gamma = 0$) and approaches 1 for a well-guiding state (β approaches $n_1 k$). With b, β is approximately expressed by

$$\beta = [1 + 2b(n_1/n_2)^2 \Delta]^{1/2} n_2 k \cong (1 + b\Delta) n_2 k \tag{2.36}$$

The phase velocity v of a plane wave is expressed as $v = c/n$, where c is the light velocity and n is the refractive index in a medium. It is convenient to express the group velocity v_g as $v_g = c/n_g$, where n_g is called the group index, and this is

$$n_g = \frac{c}{v_g} = c\frac{d\beta}{d\omega} = \frac{d\beta}{dk} = \frac{d(nk)}{dk} \tag{2.37}$$

Here, $\beta = nk$ is used. The chromatic dispersion σ_c is

$$\sigma_c = \frac{d}{d\lambda}\left(\frac{1}{v_g}\right) = -\frac{2\pi c}{\lambda^2}\frac{d^2\beta}{d\omega^2} = -\frac{1}{\lambda c}k\frac{d^2\beta}{dk^2} \tag{2.38}$$

For calculation of the chromatic dispersion σ_c,

$$\frac{d\beta}{dk} = n_{g2} + n_{g2}\Delta\frac{d(bV)}{dV} \tag{2.39}$$

Here, (2.36) is used, Δ is assumed to have no wavelength dependence, and the following equation is assumed:

$$n_{g2}\Delta \cong n_2\Delta \tag{2.40}$$

By differentiating (2.39) and using (2.38), the following equations are obtained:

$$\sigma_c = \sigma_m + \sigma_w \tag{2.41}$$

$$\sigma_m = -\frac{1}{\lambda_c}k\frac{dn_{g2}}{dk} = \frac{1}{c}\frac{dn_{g2}}{d\lambda} \tag{2.42}$$

$$\sigma_w = -\frac{2\pi\Delta}{\lambda^2}\frac{dn_{g2}}{d\omega}\frac{d(bV)}{dV} - \frac{n_{g2}\Delta}{\lambda_c}V\frac{d^2(bV)}{dV^2} \tag{2.43}$$

The derivations of $d(bV)/dV$ and $d^2(bV)/dV^2$ are obtained from the eigenvalue equation (2.15). σ_m and σ_w are the material and waveguide dispersions, respectively.

After propagating a fiber with length L, the impulse light with a spectral width $\delta\lambda$ broadens to a light with a width $\delta\tau$.

$$\delta\tau = (\sigma_m + \sigma_w)\delta\lambda \qquad (2.44)$$

Although both material and waveguide dispersions are important, the value of material dispersion is larger than that of waveguide dispersion in ordinary single-mode fibers. Therefore, the zero-dispersion wavelength λ_0 is mainly determined by material dispersion. For ordinary SI fibers, the 1.3-μm wavelength is the zero-dispersion wavelength λ_0. Waveguide dispersion is small but it can move λ_0 slightly. A fiber with $\lambda_0 = 1.55$ μm is a dispersion-shifted fiber. Dispersion-flatted fibers are also developed to realize a fiber with a low dispersion value in the 1.3- to 1.6-μm wavelength range. Several refractive-index profiles, which are different from an ordinary SI profile, have been proposed for dispersion-sifted and dispersion-flatted fibers. Typical proposed profiles and dispersions for these fibers are shown in Figures 2.3 and 2.4, respectively. The dispersions shown in Figure 2.4 include both material and waveguide dispersions. The dispersion value is important in the design of optical transmission systems because it determines the possible bit rate in the case of digital

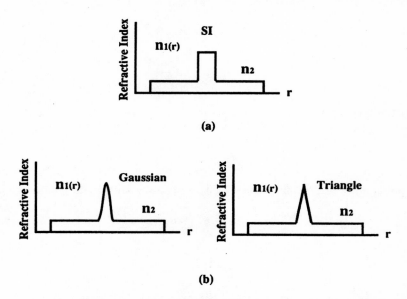

Figure 2.3 Types of single-mode fibers: (a) ordinary single-mode fiber; (b) dispersion-shifted single-mode fiber; (c) dispersion-flatted single-mode fiber.

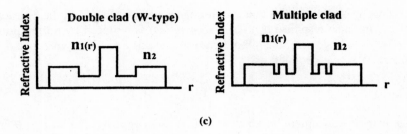

(c)

Figure 2.3 (Continued)

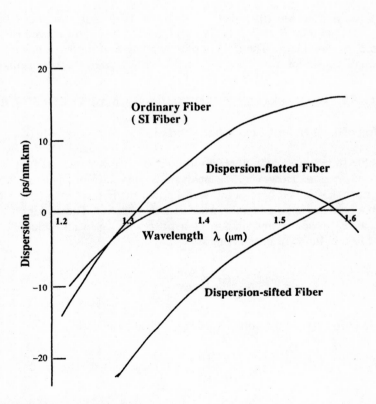

Figure 2.4 Dispersion of single-mode fibers.

transmission. It is well known that the frequency response $H(f)$ is the Fourier transform of the impulse response $h(t)$. In our case, $h(t)$ is replaced by a light pulse with a width $\delta\tau$. By considering the simple case of a small $\delta\tau$ and a symmetric $h(t)$, a 3-dB bandwidth f_{3dB} is obtained as

$$f_{3dB} = 1/(2\pi\sigma_t) \tag{2.45}$$

where σ_t is the mean square width of $h(t)$. Roughly speaking, the bit rate B is nearly equal to f_{3dB}. Since $\delta\tau$ is for a unit length, the bit rate B is roughly estimated by the following equation for fibers with length L. Here, $h(t)$ is assumed to have a rectangular shape.

$$BL \approx \frac{0.55}{(\sigma_m + \sigma_w)\delta\lambda} \tag{2.46}$$

For example, if a 10-km fiber with $\sigma_m + \sigma_w = 10$ ps/km nm and a light source with $\delta\lambda = 5$ nm, then $B \approx 1.1$ Gbps. With a narrower light source of $\delta\lambda = 0.05$ nm, then $B \approx 110$ Gbps. These examples indicate that narrow-linewidth lasers as well as small-dispersion fibers are required for high-speed fiber transmission.

2.3 TRANSMISSION CHARACTERISTICS OF MULTIMODE FIBERS

2.3.1 Mode Number and Loss Characteristics

It is possible in principle to analyze a multimode fiber with the method described in the previous section. However, the analysis is very difficult because of the large mode number. Explained here is the well-known method of analyzing a multimode fiber with a core having a general refractive-index profile. The scalar wave equation for E_x in an LP mode representation of a fiber with an infinite cladding region is expressed in a cylindrical coordinate as

$$E_x = f(r) \begin{Bmatrix} \sin v\Phi \\ \cos v\Phi \end{Bmatrix} e^{j(\omega t - \beta z)} \tag{2.47}$$

And the function $f(r)$ is the solution of the following equation.

$$\frac{d^2 f(r)}{dr^2} + \frac{1}{r}\frac{df(r)}{dr} + P(\beta,r)f(r) = 0 \tag{2.48}$$

$$P(\beta,r) = n^2 k^2 - \beta^2 - v^2/r^2 \tag{2.49}$$

where the refractive index n is

$$n = \begin{Bmatrix} n_1(r) & (0 \le r \le a) \\ n_2 & (r \ge a) \end{Bmatrix} \tag{2.50}$$

and $n_1(r)$ represents the general core profile which includes GI and SI profiles. The refractive index n_2 of a cladding is constant. In the core region, $f(r)$ is obtained using the Wentzel-Kramers-Brillouin (WKB) method [14,15]. Near the turning points r_1 and r_2, which are the solution of $P(\beta,r) = 0$, $f(r)$ is expressed in terms of Airy functions [7]. The following equations for $f(r)$ are obtained with the transformations $r = a \exp(\omega)$, $r_1 = a \exp(\omega_1)$, and $r_2 = a \exp(\omega_2)$:

$$f_1(r) = c_1(-Q)^{-1/4} \exp\left[-\int_\omega^{\omega_1} (-Q)^{1/2} \, d\omega\right] \qquad (0 \le r < r_1) \tag{2.51}$$

$$f_2(r) = 2(\pi)^{1/2} c_1(\xi/Q)^{1/4} A_i(-\xi) \qquad (r \approx r_1) \tag{2.52}$$

$$f_3(r) = 2c_1(Q)^{-1/4} \sin\left[\int_{\omega_1}^\omega (Q)^{1/2} \, d\omega + \frac{\pi}{4}\right] \qquad (r_1 < r < r_2) \tag{2.53}$$

$$f_4(r) = (\eta/Q)^{1/4} \left[c_2 A_i(-\eta) + c_3 B_i(-\eta)\right] \qquad (r \approx r_2) \tag{2.54}$$

$$f_5(r) = (4\pi)^{-1/2}(-Q)^{-1/4} \left\{c_2 \exp\left[-\int_{\omega_2}^\omega (-Q)^{1/2} \, d\omega\right] + 2c_3 \exp\left[\int_{\omega_2}^\omega (-Q)^{1/2} \, d\omega\right]\right\}$$

$$(r_2 < r \le a) \tag{2.55}$$

$$f_6(r) = c_4 H_\nu^{(1)}(j\gamma r) \qquad (r \ge a) \tag{2.56}$$

where

$$\xi = \left[\frac{3}{2} \int_{\omega_1}^\omega (Q)^{1/2} \, d\omega\right]^{2/3} = -\left[\frac{3}{2} \int_\omega^{\omega_1} (-Q)^{1/2} \, d\omega\right]^{2/3} \tag{2.57}$$

$$\eta = \left[\frac{3}{2} \int_\omega^{\omega_2} (Q)^{1/2} \, d\omega\right]^{2/3} = -\left[\frac{3}{2} \int_{\omega_2}^\omega (-Q)^{1/2} \, d\omega\right]^{2/3} \tag{2.58}$$

$$Q = r^2 P(\beta,r) \tag{2.59}$$

$$\gamma^2 = \beta^2 - n_2^2 k^2 \tag{2.60}$$

where

$f_1(r), f_2(r), \ldots, f_6(r)$ = function $f(r)$ for different regions of r;
$C_1, C_2, \ldots, C_4$ = constants;
A_i, B_i = two linearly independent Airy functions.

The other field components can be derived from E_x. Using boundary conditions at $r = r_2$, $r = a$, and $r = b$, the following result is obtained.

$$\int_{r_1}^{r_2} [P(\beta,r)]^{1/2} \, dr = \left(\mu - \frac{1}{2}\right)\pi \qquad (\mu = 1, 2, 3, \ldots) \qquad (2.61)$$

This equation corresponds to the eigenvalue equation, and β for the $LP_{\nu\mu}$ mode is given by solving it. For a given β range such as $\beta_1 < \beta < \beta_{max}$, number of possible pairs (ν,μ) is given by (2.61). Therefore, the mode number can be calculated by the equation. Below, the general core profile $n_1(r)$ is restricted to the α index profile of (1.3). The mode number $s(\beta)$ is expressed by [16]

$$s(\beta) = \frac{\alpha}{\alpha + 2} \frac{V^2}{2} \left[\frac{a^2(k^2 n_0^2 - \beta^2)}{V^2}\right]^{2/\alpha+1} \qquad (2.62)$$

where $s(\beta)$ represents the number of modes having a propagation constant larger than β. Equation (2.62) is obtained by counting the number of modes using (2.61), using the approximation that summation is replaced by integration. Fourfold mode degeneracy is considered. The total number of modes N is obtained with $\beta = \beta_{min}$, where $\beta_{min} = n_2 k$. That is,

$$N = \frac{\alpha}{\alpha + 2} \frac{V^2}{2} \qquad (2.63)$$

Examples of the total number of modes of SI and GI ($\alpha = 2$) fibers with parameters of core diameter $2a = 50$ μm, $n_0 = 1.46$, $\Delta = 1\%$, and $\lambda = 1$ μm are

$$N = \frac{V^2}{2} \approx 526 \quad (SI)$$

$$N = \frac{V^2}{4} \approx 263 \quad (GI)$$

These examples show that the total number of modes is very large in a multimode fiber. By rewriting (2.62) and using N, β is

$$\beta = n_0 k \left[1 - 2\Delta \left(\frac{s}{N} \right)^{\alpha/\alpha+2} \right]^{1/2} \tag{2.64}$$

The physical image of an optical wave (ray) in a multimode fiber is shown in Figure 2.5. The propagation angle θ as indicated in this figure is related to β:

$$\beta = n_0 k \cos\theta \tag{2.65}$$

The range of β is

$$n_2 k < \beta < n_0 k \tag{2.66}$$

The higher modes have smaller β values and the propagation angle θ is larger, as indicated by (2.65). The ray is reflected by a core-cladding interface, and more reflections take place for rays with larger θ values. Due to the core-cladding imperfection, higher modes have larger propagation loss. This is also explained by field theory: the field of lower modes is strongly confined in the core region. On the other hand, the field of higher modes leaks into the cladding. Therefore, higher modes tend to have larger propagation loss. This also holds in the case of a fiber with a lossy third layer (an outer layer of the cladding) [4]. Due to the above circumstances, the measured loss of a multimode fiber depends on the mode power distribution. The loss of a multimode fiber is usually measured by using a steady-state mode power distribution, which is defined as the power distribution in a sufficiently long fiber length. Actually, the output of either a fiber of a few kilometers wound on a small drum or a mode scrambler is used for the steady-state mode power distribution in a measurement. Not only the loss but also the baseband response depends on the mode power distribution in the case of multimode fibers. This will be discussed in the next section. These mode power distribution dependencies are very important for discussing or measuring passive optical components. The realized multimode fiber

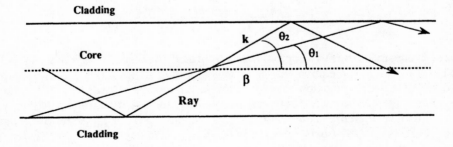

Figure 2.5 Ray trace in a multimode fiber.

loss as a function of wavelength is similar to the single-mode fiber loss shown in Figure 2.1.

2.3.2 Modal Dispersion and Baseband Characteristics

The group delay τ for a fiber with an α profile is obtained by (2.9) and (2.64). The results are

$$\tau(m) = \frac{n_{g0}}{c}\left[1 + \Delta\left(\frac{\alpha - 2 - y}{\alpha + 2}\right)\left(\frac{s}{N}\right)^{\frac{\alpha}{\alpha+2}} + \Delta^2\left(\frac{3\alpha - 2 - 2y}{2\alpha + 4}\right)\left(\frac{s}{N}\right)^{\frac{2\alpha}{\alpha+2}}\right] \quad (2.67)$$

$$n_{g0} = n_0 - \lambda\frac{dn_0}{d\lambda} \quad (2.68)$$

$$y = -\frac{2n_0}{n_{g0}}\frac{\lambda}{\Delta}\frac{d\Delta}{d\lambda} \quad (2.69)$$

where s and N are the mode number and the total mode number, respectively and n_g is the group index. It can clearly be seen from these equations that the group delay difference among modes can be decreased when $\alpha = 2 + y$. For multimode fibers with $\alpha = 2 + y$, the group delay difference is proportional to Δ^2 and results in a small value (i.e., fibers with wide bandwidth). The parameter y is related to the wavelength dependence of the refractive index [17]. Therefore, the optimum α value, $\alpha_{opt} = 2 + y$, depends on the wavelength, fiber parameters, and materials. For a 1.3-μm wavelength, α_{opt} is 1.9 for a GeO_2 dopant fiber with $\Delta = 1\%$.

The physical image of wide bandwidth for multimode fibers with $\alpha \approx 2$ (GI multimode fibers) is explained in Figure 2.5, which is drawn for an SI fiber ($\alpha \rightarrow \infty$) for simplicity. In GI multimode fibers, the rays follow curved paths, not straight paths. The propagation time (delay) t for a fiber length L is

$$t = \frac{L}{\overline{v_g}\cos\theta} = \frac{L\overline{n_g(r)}}{c\cos\theta} \approx \frac{L\overline{n(r)}}{c\cos\theta} \quad (2.70)$$

Here, group index $n_g(r)$ is approximately replaced by $n(r)$. The symbols $\overline{v_g}$, $\overline{n_g(r)}$, and $\overline{n(r)}$ represent the equivalent (or averaged) values of v_g, $n_g(r)$, and $n(r)$, respectively. Since the propagation angle θ of the lower modes is small, the lower modes travel near the core center. Therefore, $\overline{n(r)}$ is $n(r_1)$ for the lower modes, where r_1 is the radius near the core center. For the higher modes, $\overline{n(r)} = n(r_2)$ and $0 < r_1 < r_2 < a$. For $\alpha \rightarrow \infty$ (SI fiber), $n(r_1) = n(r_2) = n_1 = $ constant. The value of $\cos\theta$ is larger for the lower modes. Therefore, $t_{lower} < t_{higher}$, where t_{lower} and t_{higher} are the propagation times for the lower and higher modes, respectively. For the small value of α, $n(r_1) > n(r_2)$. When the ratio of $n(r)/\cos\theta$ is equal for the lower and higher modes, equal propagation time is realized. This is the case in $\alpha \approx 2$. The lower modes propagate near the core center where the refractive index is high. Because of the high refractive index, the propagation speed is slow compared to that of the higher modes. However, the propagation distance for the lower modes is short. With slow speed and short distance for the lower modes and high speed and long distance for the higher modes, nearly the same delay is realized. When $\alpha > \alpha_{opt}$, the speed of the lower modes is faster than that of the higher modes. When $\alpha < \alpha_{opt}$, the situation is reversed (Figure 2.6).

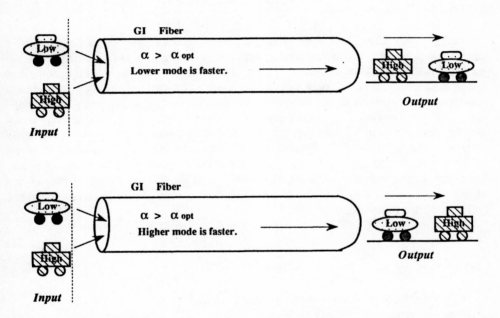

Figure 2.6 Group velocity difference of modes in a multimode fiber. Each car corresponds to a mode in a fiber, and the labels "low" and "high" indicate the lower and higher modes, respectively.

2.3.3 Mode Conversion

Modes propagating in a multimode fiber convert to other propagating modes or radiation modes, a process called mode conversion. Mode conversion originates in several factors, listed in Table 2.1. Even in a straight fiber, fluctuations of a core-cladding interface and a refractive-index profile $n_1(r)$ cause mode conversion. Mode

Table 2.1
Causes of Mode Conversion

Category	Phenomenon
Fiber itself	(1) Core-cladding interface fluctuation
	(2) Refractive index profile fluctuation
Environment and component	(3) Bent
	(4) Side pressure
	(5) Vibration
	(6) Connection
	(7) Others (bent in waveguide, core deformation, etc.)

conversion also takes place when fibers are bent, pressed, and vibrated. In passive optical components, mode conversion occurs in connections between fibers or between a fiber and a component. In components themselves, mode conversion takes place in several situations, such as bent waveguides in a waveguide-type component and core deformations in a fiber-type component. The basic equation for a mode conversion is the coupled power equation. Although this equation is derived based on the fluctuations in a fiber [18], here we derive the equation based on the linearity of power. According to S.D. Personick [19], fibers act as a linear system in intensity when the following two conditions are satisfied: (1) mode transfer function is stationary; (2) mode transfer functions are uncorrelated. We then make use of Personick's result to derive the coupled power equation. The linearity of power results in the validity of the following expression:

$$[P_i]_{\text{out}} = [C_{ij}][P_j]_{\text{in}} \tag{2.71}$$

where

$[C_{ij}]$ = matrix of mode conversion coefficients;
$[P_j]_{\text{in}}$ = input mode power distribution;
$[P_i]_{\text{out}}$ = output mode power distribution.

When the locations of input and output are taken to be z and $z + \Delta z$, (2.71) for each mode can be written as

$$P_i(z + \Delta z) = \sum_j C_{ij} P_j(z) \tag{2.72}$$

The left-hand side of this equation is in a Taylor expansion; we take the first term and then

$$P_i(z) + \frac{dP_i}{dz} \Delta z = \sum_j C_{ij} P_j(z) \qquad (2.73)$$

$P_i(z)$ is also expressed in the following form:

$$P_i(z) = \sum_j C_{ji} P_i(z) + L_i P_i(z) \qquad (2.74)$$

where L_i is the loss factor of mode i. Equation (2.74) shows that mode i converts to mode j and radiation modes. The converted power to mode j is the summation of all propagation modes including itself. The power converted to radiation modes is expressed in the loss factor. By substituting (2.74) into (2.73) with $\Delta z = 1$, we obtain

$$\frac{dP_i}{dz} = -2\alpha_i P_i + \sum_j h_{ij} P_j - \sum_j h_{ji} P_i \qquad (2.75)$$

Here, h_{ij} and $2\alpha_i$ represent the C_{ij} per unit length and the loss coefficient per unit length, respectively. By considering the time dependence, the left-hand side of (2.75) is

$$\frac{dP(z,t)}{dz} = \frac{\partial P(z,t)}{\partial z} + \frac{\partial P(z,t)}{\partial t} \frac{\partial t}{\partial z} = \frac{\partial P}{\partial z} + \frac{1}{v} \frac{\partial P}{\partial t} \qquad (2.76)$$

Then (2.75) results in

$$\frac{\partial P_i}{\partial z} + \frac{1}{v_i} \frac{\partial P_i}{\partial t} = \left(-2\alpha_i - \sum_j h_{ji} \right) P_i + \sum_j h_{ij} P_j \qquad (2.77)$$

Equation (2.77) is the differential form and is called the coupled power equation. The steady-state mode power distribution is obtained for an infinite z length when h_{ij} is length-independent. In the steady state, output mode power distribution is independent of the input mode power distribution. It only depends on the mode conversion h_{ij} in a fiber. We show that (2.71) is equivalent to (2.77). When we treat passive components such as connectors, (2.71) is better than (2.77) in most cases. The input power distribution changes according to the mode conversion coefficient matrix in the case of (2.71).

The effects of mode conversions are listed in Table 2.2. When the guided modes couple to other guided modes, this type of mode conversion is called mode mixing. Mode mixing averages the group velocities of guided modes. Radiation loss takes place when the guided modes couple to radiation modes. A representation of mode conversion is shown in Figure 2.7. Mode mixing decreases the group velocity

Table 2.2
Effects of Mode Conversion

Category	Effects
Coupling to guided modes (mode mixing)	Averaging of group velocities
Coupling to radiation modes (radiation loss)	Loss of guided mode power

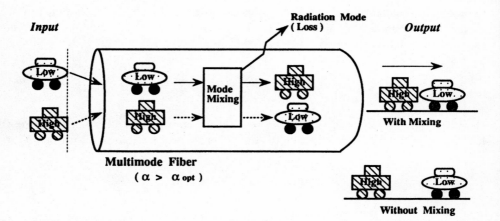

Figure 2.7 Effect of mode conversion in a multimode fiber. (Represents the case of $\alpha > \alpha_{opt}$, where the higher mode is faster than the lower mode.)

difference, and this results in realizing a wider bandwidth in most cases. However, this is not true for the spliced or connected fibers, when one fiber with $\alpha > \alpha_{opt}$ and another fiber with $\alpha < \alpha_{opt}$. This situation will be explained in Chapter 3.

2.4 MECHANICAL CHARACTERISTICS

Mechanical characteristics are important for handling or working on fibers. The tensile strength of silica fibers without a coating is 20 to ~30 kg/mm², and it improves if the fibers have a plastic coating. With the coating, the surface of the fibers is protected from flaws or cracks. With a coating of 10-μm thickness, the strength is 500 to 600 kg/mm², corresponding to 5.5 to 6.6 kg for fibers with an outer diameter of 125 μm. Monocoated fibers with primary and secondary coatings have a strength of about 7 kg, strong enough for ordinary handling. Fiber ribbon is about n times stronger than the monocoated fiber, where n is the fiber number in a fiber ribbon.

A fiber with flaws or cracks becomes weak. The stress σ_f causing failure is related to the flaw size by the following equation.

$$\sigma_f = \frac{K_{IC}}{YC_c^{1/2}} \tag{2.78}$$

where K_{IC}, C_C, and Y are the critical stress intensity factor (or the fracture toughness), the critical flaw size, and the factor describing flaw geometry, respectively. This equation indicates that fibers with a smaller flaw size are stronger. An example in [20] indicates that the flaw sizes on the fiber surface must be smaller than about 0.4 μm in order for the fiber to survive an external stress of 1 GN/m^2. It is important to avoid large flaws in the fabrication of fiber components. Since cracks or flaws are created in the fusion splice process, the spliced portion must be protected to improve its strength. It is known that a fiber with cracks becomes weak if it is underwater or in water vapor. It is believed that cracks grow in these environments. Growth speed v is often assumed to be

$$v = AK_I^n \tag{2.79}$$

where A, K_I, and n are the material constant, the stress intensity factor, and the stress corrosion susceptibility factor, respectively. The factor n is often used to describe fiber fatigue.

REFERENCES

[1] Okoshi, T., Optical Fibers, Academic Press, 1982.
[2] Yokota, H., H. Kanamori, Y. Ishiguro, G. Tnaka, H. Takada, M. Watanabe, S. Suzuki, K. Yano, M. Hoshikawa, and H. Shimba, "Ultra-Low-Loss Pure-Silica-Core Single-Mode Fiber and Transmission Experiment," Conf. on Optical Fiber Communication (OFC'86), PD3, Atlanta, 1986.
[3] Gloge, D., "Weakly Guiding Fibers," Appl. Opt., Vol. 10, 1971, p. 2252.
[4] Kashima, N. and N. Uchida, "Transmission Characteristics of Graded-Index Optical Fibers With a Lossy Outer Layer," Appl. Opt., Vol. 17, 1978, p. 1199.
[5] Marcuse, D., Theory of Dielectric Optical Waveguides, Academic Press, 1974.
[6] Adams, M. J., An Introduction to Optical Waveguides, John Wiley & Sons, 1981.
[7] Abramowitz, M. and I. A. Stegun, Handbook of Mathematical Functions, Dover, 1972.
[8] Marcuse, D., "Gaussian Approximation of the Fundamental Modes of Graded-Index Fibers," J. Opt. Soc. Am., Vol. 68, 1978, p. 103.
[9] Marcuse, D., "Loss Analysis of Single-Mode Fiber Splices," Bell Syst. Tech. J., Vol. 56, 1977, p. 703.
[10] Petermann, K., "Microbending Loss in Monomode Fibers," Electron. Lett., Vol. 12, 1976, p. 107.
[11] Petermann, K., "Constraints for Fundamental-Mode Spot Size for Broadband Dispersion-Compensated Single-Mode Fibers," Electron. Lett., Vol. 19, 1983, p. 712.

[12] Pask, C., "Physical Interpretation of Petermann's Strange Spot Size for Single-Mode Fibers," Electron. Lett., Vol. 20, 1984, p. 144.

[13] Ohashi, M., N. Kuwaki, and N. Uesugi, "Suitable Definition of Mode Field Diameter in View of Splice Loss Evaluation," IEEE J. Lightwave Technol., Vol. LT-5, 1987, p. 1676.

[14] Shiff, L., Quantum Mechanics, McGraw-Hill, 1968.

[15] Petermann, K., "The Mode Attenuation in General Graded-Core Multimode Fibers," A.E.Ü, Vol. 29, 1975, p. 345.

[16] Gloge, D. and E. A. J. Marcatili, "Multimode Theory of Graded-Core Fibers," Bell Syst. Tech. J., Vol. 52, 1973, p. 1563.

[17] Olshansky, R., "Mode Coupling Effects in Graded-Index Optical Fibers," Appl. Opt., Vol. 14, 1975, p. 935.

[18] Marcuse, D., "Derivation of Coupled Power Equations," Bell Syst. Tech. J., Vol. 51, 1972, p. 229.

[19] Personick, S. D., "Time Dispersion in Dielectric Waveguides," Bell Syst. Tech. J., Vol. 50, 1971, p. 843.

[20] Paek, U. C., "High-Speed High-Strength Fiber Drawing," IEEE J. Lightwave Technol., Vol. LT-4, 1986, p. 1048.

Chapter 3
Transmission Characteristics of Passive Optical Components

There are many passive optical components used for fiber transmission, and the transmission characteristics depend both on fibers and optical components. Detailed characteristics particular to the individual component will be discussed in Parts II and III of this book. Because the simple connection is the basic function of passive optical components for fiber transmission, we discuss the transmission characteristics of splices and connectors in this chapter. Discussed characteristics resulting from a connection are connection loss and reflection for both types of fibers and a mode conversion for multimode fibers. The discussion of these characteristics will be helpful for understanding passive optical components other than splices and connectors.

3.1 CONNECTION LOSS

Several factors influence fiber connection loss, and they are listed in Table 3.1. These factors can be divided into two groups: intrinsic and extrinsic factors. Intrinsic factors cause connection loss even in a perfect connection state, and they originate in the parameter mismatch of two mated fibers. Extrinsic factors originate in imperfect connection states. Outer-diameter mismatch causes connection loss only for the connection method using the outer diameter. It does not matter for a method that does not use the outer diameter, such as several core alignment methods. There is no loss caused by the longitudinal separation in a fusion splice, because two mated fibers are fused together without a gap. The refractive index at the connection portion is changed by the index-matching material or air between two fiber facets in the case of connectors and mechanical and adhesive splices, and by the heat-induced deviation from the original index profile in the case of fusion splices. The heat-induced deviation may be caused by the movement of softened silica glass from pressing and

Table 3.1
Factors of Connection Loss

Factor	Single-Mode Fibers	GI Multimode Fibers
Intrinsic factor	MFD mismatch	Core diameter mismatch Δ mismatch α mismatch
	Outer-diameter mismatch*	
Extrinsic factor	Lateral offset Longitudinal separation Axial tilt Facet quality Refractive-index change	

*Only for the method using outer diameter.

by a diffusion of dopants. In an ordinary connection method, the lateral offset and axial tilt are the two major extrinsic factors.

3.1.1 Connection Loss in Single-Mode Fibers

Connection loss can be evaluated by using the field of fibers. We express the electric fields of two mated fibers (input and output fibers) using E_{in} and E_{out}, respectively. Input field E_{in} can be expanded by mode e_i of the output fiber:

$$E_{in} = \sum_i \xi_i e_i \qquad (3.1)$$

where e_i includes the radiation modes and ξ_i represents the expansion coefficient. e_0 is the fundamental mode, and we assume that this is the only guiding mode in single-mode fibers (we are only considering the single-mode region). When $i > 0$, e_i symbolizes the radiation modes. By multiplying $(1/2)h_0^*$ and integrating the fiber cross-sectional plane, we obtain

$$\frac{1}{2}\int_s E_{in} h_0^* \, dS = \frac{1}{2}\xi_0 \int_s e_0 h_0^* \, dS + \frac{1}{2}\sum_{i>0} \xi_i \int_s e_i h_0^* \, dS \qquad (3.2)$$

The second term of the right-hand side of this equation becomes zero because of the mode orthogonality, and this results in

$$\xi_0 = \frac{\int_s E_{in} h_0^* \, dS}{\int_s e_0 h_0^* \, dS} \tag{3.3}$$

The power of the guiding mode in the input and output fibers is

$$P_{in} = \frac{1}{2} \int_s E_{in} H_{in}^* \, dS \tag{3.4}$$

$$P_{out} = \frac{1}{2} |\xi_0|^2 \int_s e_0 h_0^* \, dS \tag{3.5}$$

Then the power transmission coefficient T is

$$T = \frac{P_{out}}{P_{in}} = \frac{\left| \int_s E_{in} h_0^* \, dS \right|^2 \int_s e_0 h_0^* \, dS}{\left| \int_s e_0 h_0^* \, dS \right|^2 \int_s E_{in} H_{in}^* \, dS} = \frac{\left| \int_s E_{in} h_0^* \, dS \right|^2}{\int_s E_{in} H_{in}^* \, dS \int_s e_0 h_0^* \, dS} \tag{3.6}$$

For simplicity, we replace $E_{in} = f(x,y)$ and $E_{out} = g(x,y)$ and use the fact that the magnetic field H has a linear relationship with the electric field E. Then (3.6) is expressed as

$$T = \frac{\left| \int_s fg \, dS \right|^2}{\int_s f^2 \, dS \int_s g^2 \, dS} \tag{3.7}$$

where $f(x,y)$ and $g(x,y)$ are assumed to be real functions.

The fundamental mode of single-mode fiber is approximately expressed by the Gaussian function as discussed in Chapter 2. By using (3.7), T can be calculated for the three major factors shown in Figure 3.1. The results are shown below [1].

1. Lateral offset (offset is d):

$$T = \left(\frac{2\omega_1\omega_2}{\omega_1^2 + \omega_2^2} \right)^2 \exp\left(-\frac{2d^2}{\omega_1^2 + \omega_2^2} \right). \tag{3.8}$$

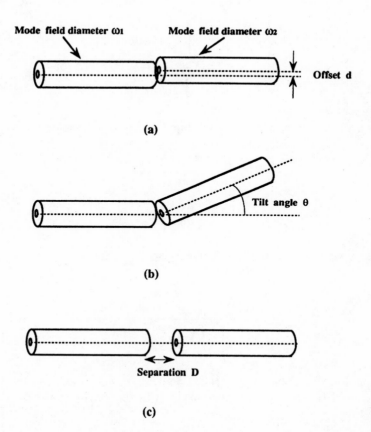

Figure 3.1 Three major connection loss factors: (a) lateral offset; (b) axial tilt; (c) end separation.

2. Axial tilt (tilt angle is θ):

$$T = \left(\frac{2\omega_1\omega_2}{\omega_1^2 + \omega_2^2}\right)^2 \exp\left(-\frac{2(\pi n_2\omega_1\omega_2\theta)^2}{(\omega_1^2 + \omega_2^2)\lambda^2}\right) \tag{3.9}$$

3. Longitudinal separation (separation is d_z):

$$T = \frac{4\left[4Z^2 + \dfrac{\omega_1^2}{\omega_2^2}\right]}{\left[4Z^2 + \dfrac{\omega_1^2 + \omega_2^2}{\omega_2^2}\right] + 4Z^2\dfrac{\omega_2^2}{\omega_1^2}} \tag{3.10}$$

$$Z \equiv \frac{d_z \lambda}{2\pi n_2 \omega_1 \omega_2} \tag{3.11}$$

These equations contain both intrinsic and extrinsic loss factors, except for facet quality and refractive-index change. Outer-diameter mismatch can be interpreted as a lateral offset in the connection methods using the outer diameter. Among the three equations, only (3.8) is derived here. Two mode fields of mated single-mode fibers, whose MFDs are ω_1 and ω_2, with lateral offset d are expressed as follows:

$$f(x, y) = A \exp\left(-\frac{x^2 + y^2}{\omega_1^2}\right) \tag{3.12}$$

$$g(x, y) = B \exp\left(-\frac{(x + d)^2 + y^2}{\omega_2^2}\right) \tag{3.13}$$

where A and B are constants.

$$\int f^2 \, dS = \frac{\pi}{2} A^2 \omega_1^2 \text{ and } \int g^2 \, dS = \frac{\pi}{2} B^2 \omega_2^2 \tag{3.14}$$

$$\int f(x,y)g(x,y) \, dS = AB \int_{-\infty}^{\infty} \exp\left[-\left(\frac{1}{\omega_1^2} + \frac{1}{\omega_2^2}\right)y^2\right] dy \int_{-\infty}^{\infty} \exp\left[-\left(\frac{x^2}{\omega_1^2} + \frac{(x+d)^2}{\omega_2^2}\right)\right] dx$$

$$= AB \sqrt{\frac{\pi\omega_1^2\omega_2^2}{\omega_1^2 + \omega_2^2}} \sqrt{\frac{\pi\omega_1^2\omega_2^2}{\omega_1^2 + \omega_2^2}} \exp\left(\frac{-d^2}{\omega_1^2 + \omega_2^2}\right) \tag{3.15}$$

$$= AB\left(\frac{\pi\omega_1^2\omega_2^2}{\omega_1^2 + \omega_2^2}\right) \exp\left(\frac{-d^2}{\omega_1^2 + \omega_2^2}\right)$$

With (3.14) and (3.15), we obtain (3.8). For fibers with the same mode field ω, connection loss L corresponding to (3.8) is simply expressed as

$$L = -10 \log T \approx 4.34 \left(\frac{d}{\omega}\right)^2 \quad \text{(dB)} \tag{3.16}$$

For lateral displacement $d = 1 \ \mu m$, L is calculated to be 0.27 dB for $\omega = 4 \ \mu m$. To keep the loss below 0.25 dB, lateral displacement must be under 1 μm (submicron). Practical loss factors are not only lateral displacement, but also angular displacement and fiber facet quality. The submicrometer accuracy is required in an outside plant (a very hard environment when compared to room or telephone office environment) and is also required in the long term for practical splices or connectors used in the outside plant.

Using (3.8) through (3.11) and (2.28), connection losses of an SI single-mode fiber as a function of wavelength λ can be calculated and the results are shown in Figure 3.2(a–c). Here, we assume the case of two identical fibers with parameters of a = 4 μm and V = 2.405 for λ = 1.2 μm. We also assume that Δ is wavelength-independent. For longer wavelengths, connection loss becomes smaller with the parameters used in these calculations. This is the result of the larger MFD for longer wavelengths.

3.1.2 Connection Loss in Multimode Fibers

Connection loss in multimode fibers can be evaluated by using (3.7) for each guiding mode. The connection loss for a multimode fiber is the summation of connection loss considering the mode power distribution. That is,

$$L_{all} = \sum_i L(i)P(i) \tag{3.17}$$

where L_{all}, $L(i)$, and $P(i)$ are the connection loss, each connection loss for mode i, and the mode power for mode i, respectively. Because the connection loss is weighted by the mode power, the connection loss for a multimode fiber strongly depends on the mode power distribution. For example, measuring splice loss by using a light-emitting diode (LED) is quite different from measuring with a laser diode. A light source like an LED tends to excite all modes in a multimode fiber, and the mode power distribution is similar to the uniform distribution. In this case, splice loss is large when compared to the case of steady-state mode power distribution. The laser like a laser diode tends to excite specific modes. When lower modes are excited by a laser diode, the splice loss is smaller than in the case of an LED excitation. One example is that a given splice loss is measured to be 0.2 dB with a laser diode and 0.4 dB with an LED. Many theoretical formulas for calculating connection loss are proposed, and some of them are listed in [2–5]. The formulas are developed by considering the local mode volume (or local numerical aperture) at the overlapped area of mated fibers. Generally, connection loss of a multimode fiber is discussed or measured in the steady-state (or nearly-steady-state) mode power distribution, because splices or connectors are mainly used in relatively long fiber chains for fiber transmission. Although the proposed formulas are useful, the appropriate mode power distribution must be assumed for evaluating the connection loss in the steady state. Instead of the theoretical formulas, we present the following simple formulas for evaluating connection loss in the steady state [6].

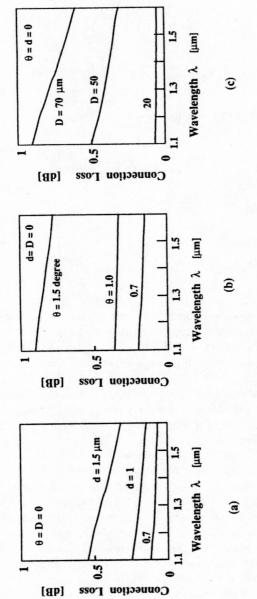

Figure 3.2 Connection loss in single-mode fibers: (a) lateral offset; (b) axial tilt; (c) end separation.

1. Lateral offset d:

$$L = 3 \frac{d}{2a} \quad \text{(dB)} \tag{3.18}$$

2. Core diameter difference in the case of $a_1 \geq a_2$:

$$L = 4.5 \left(1 - \frac{a_2}{a_1} \right) \quad \text{(dB)} \tag{3.19}$$

3. Difference in refractive index in the case of $\Delta_1 > \Delta_2$:

$$L = 1.6 \left(1 - \frac{\Delta_2}{\Delta_1} \right) \quad \text{(dB)} \tag{3.20}$$

where $2a$ and Δ are the core diameter and the relative refractive-index difference, respectively. These are obtained experimentally and are useful for regions of relatively small connection loss ($L < 0.5$ dB). The fibers used in the experiments are GI multimode fibers ($a \approx 2$) with parameters of $2a = 50$ μm, $\Delta = 1\%$, and outer diameter = 125 μm.

Using these equations for multimode fibers, splice loss caused by fiber parameter differences can be calculated. Figure 3.3 shows the difference between measured and calculated splice losses as a function of transmission length Z_t. Z_t is defined as the length from the light source, which is an LED with a 0.85-μm wavelength in this experiment. The fibers with parameters of $2a = 50$ μm, $\Delta = 1\%$, and outer diameter = 125 μm for standard values are used. Fibers are in 10-fiber cables installed in the field. Splices are made with the fusion splice method. The measured splice loss is the average loss of 10 splices. The calculated splice losses are based on the steady-state mode power distribution, while the power distribution for the measured ones depends on Z_t. Since there are much higher modes in a fiber located near the light source, splice losses are measured high for short Z_t. This figure shows experimentally that connection loss for a multimode fiber depends strongly on the transmission length from a light source.

The comparison of connection losses for single-mode and multimode fibers is shown in Figure 3.4, where $\omega = 5$ μm for single-mode fibers and $2a = 50$ μm, $\Delta = 1\%$, and $\alpha = 2$ for multimode fibers are assumed in these calculations.

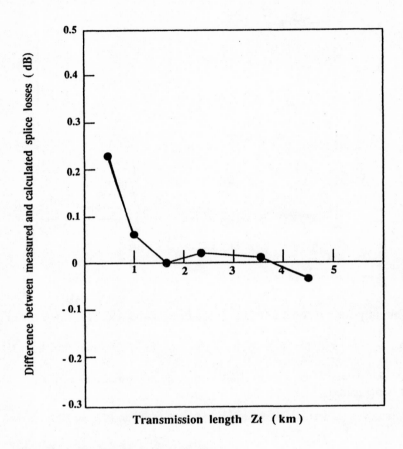

Figure 3.3 Transmission length dependence of splice loss.

3.2 MODE CONVERSION BY CONNECTION IN MULTIMODE FIBERS

3.2.1 Experimental Investigation of Mode Conversion

Mode conversions caused by splices are measured using the FFP changes before and after splicing [7]. Fibers used in the experiments are multimode fibers with $2a = 60 \ \mu$m, $\Delta = 1\%$, and $\alpha \approx 2$. The fiber length is 2 km and an LED with a wavelength of $\lambda = 0.84 \ \mu$m is used as a light source. Two types of splice methods, fusion splice and adhesive-bonded V-groove splice, are used. In the V-groove splice, index materials are used. The measuring process is:

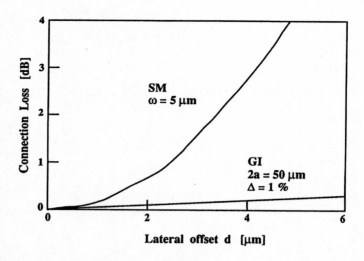

Figure 3.4 Comparison of connection loss in single-mode and multimode fibers.

1. Output power P_0 of 2-km fiber is first measured and FFP $P_0(\theta)$ for angle θ is measured.
2. A 2-km fiber is cut at a point 12 m from the output end and then fibers are spliced by either fusion or V-groove splice.
3. Output power P_1 of a spliced 2-km fiber is measured and FFP $P_1(\theta)$ for angle θ is measured.

Splice loss L, mode conversion parameter $d(\theta)$, and D are defined as

$$L = -10 \log \frac{P_1}{P_0} \qquad \text{(dB)} \tag{3.21}$$

$$d(\theta) = \frac{P_1(\theta) - P_0(\theta)}{P_0} \tag{3.22}$$

$$D = \sum_{\theta} |d(\theta)| \tag{3.23}$$

where $P_0(\theta)$ and $P_1(\theta)$ are measured for every 1° in the air. The symbol Σ_{θ} represents the summation of all θ values. Then P_0 and P_1 are expressed as

$$P_0 = \sum_{\theta} P_0(\theta) \tag{3.24}$$

$$P_1 = \sum_\theta P_1(\theta) \qquad (3.25)$$

Although strictly speaking FFP does not completely correspond to mode power distribution for GI fibers, it can be used to evaluate mode power distribution. Therefore, $d(\theta)$ and D are considered measure for evaluating mode conversion. The measured $d(\theta)$ is shown in Figure 3.5. In Figure 3.5(a,b), splice loss is taken as a parameter. Before comparing the two cases, the meaning of $d(\theta)$ is considered:

1. $d(\theta) > 0$ means that mode power flows into a given θ from other θ values by means of a splice.
2. $d(\theta) < 0$ means that power flows out somewhere (including radiation modes) by means of a splice.
3. $d(\theta) = 0$ means that there is no mode power movement to other θ values, or that the inflow and outflow are balanced.

Therefore, the following results are obtained by comparing Figure 3.5 (a,b):

1. For the lower modes, mode power movement in fusion splices is larger than that in V-groove splices.
2. For the middle modes, mode power is lost by moving to other modes in both kinds of splices.
3. For the higher modes, mode power flows in both kinds of splices. There is much inflow in fusion splices.

Figure 3.6 shows the relationship between the splice loss and the macro mode conversion parameter D for both splice methods. In this figure, solid curves a and b represent experimental values of fusion splices and V-groove splices, respectively. The dotted curve c represents theoretical values for the case where the guided modes directly couple to the radiation modes without internal mode movements among the guided modes. An approximate linear relationship between the mode conversion and splice loss is found to hold for both splicing methods when the splice loss is small. It is clear from Figure 3.6 that a splice made with the fusion splice method has a larger mode conversion than a splice made with the V-groove splice method for the same splice loss.

3.2.2 Theoretical Investigation of Mode Conversion

The difference in mode conversion between a fusion splice and a V-groove splice can be clarified by modeling the spliced part of both splices [8–10]. The model of the spliced part without lateral offset and tilt is shown in Figure 3.7 for both splices.

In a V-groove splice, index-matching materials are used between fiber ends. The refractive index of matching materials is a constant value near that of fiber. For the modeling of fusion splices, the refractive-index profile changes made by fusion

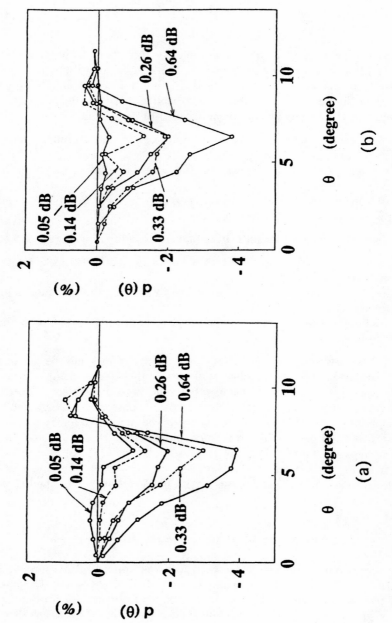

Figure 3.5 Measured $d(\theta)$: (a) $d(\theta)$ in fusion splices; (b) $d(\theta)$ in adhesive splices (*V*-groove). (After [7].)

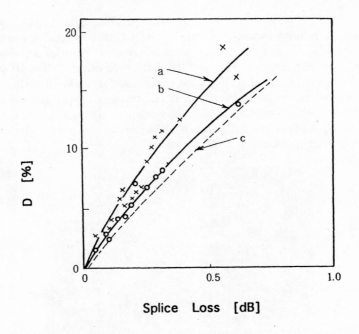

Figure 3.6 Measured relationship between splice loss and mode conversion. (After [7].)

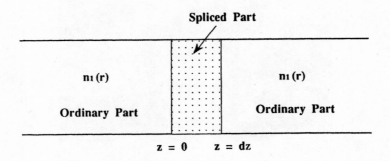

Figure 3.7 Model of spliced fibers.

splices are measured by an interference microscope. The spliced fiber perpendicular to the fiber axis is used in the experiment. Both fusion spliced part and ordinary fiber part are measured and their interference microscope pictures are shown in Figure 3.8 (a,b), respectively [8]. To show the difference clearly, SI multimode fibers are used. The circles in these pictures are core, cladding, and silica jacket, respectively. By comparing the two pictures, it is clear that the core-cladding interface is smoothed and the refractive index is also changed. It is anticipated that these changes are caused by the movement of dopants in a fiber or the movement of softened silica glass by the pressing process in a fusion splice.

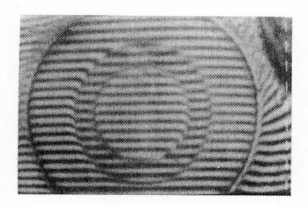

(b) Ordinary part

Figure 3.8 Interference microscope pictures for SI fibers: (a) fusion splice part; (b) ordinary part.

Considering these results, the refractive index of the spliced part is assumed to be

$$n(x,y,z) = n_s(x,y) + q(x,y,z) \quad (0 \le z \le d_z) \tag{3.26}$$

$$n_s(x,y) = n_0 \left[1 - \Delta \left(\frac{x^2 + y^2}{a^2} \right) \right] \tag{3.27}$$

where $n_s(x,y)$ and $q(x,y,z)$ are the refractive-index profile for the ordinary part and the profile derivation for the spliced part. The derivation is assumed to be small; that is,

$$|q/n_s| \ll 1 \tag{3.28}$$

The deviated length dz is assumed to be small. We consider the simple situation that two mated fibers with a square-low index profile are identical. The electric field in the splice region E is expressed by the set of $[\hat{E}^i]$, which is the electric field of the ordinary fiber part with $n_s(x,y)$.

$$E = \sum_i g_i(z) \hat{E}^i \exp(-j\beta_i z) \tag{3.29}$$

where $g_i(z)$ is a coefficient for mode i and a function of z, and β_i is the propagation constant for mode i. The symbol Σ_i represents the summation of all guided modes and the integration of all radiation modes. The following wave equation is derived from Maxwell's equation:

$$\frac{\partial^2 E}{\partial x^2} + \frac{\partial^2 E}{\partial y^2} + \frac{\partial^2 E}{\partial z^2} + k^2 n^2 E = 0 \tag{3.30}$$

By substituting (3.29) into (3.30), the following equation for $g_i(z)$ is obtained:

$$\sum_i g_i(z) \left(\frac{\partial^2 \hat{E}^i}{\partial x^2} + \frac{\partial^2 \hat{E}^i}{\partial y^2} + k^2 n^2 \hat{E}^i \right) + \sum_i \left(\frac{\partial^2 g_i}{\partial z^2} - 2j\beta_i \frac{\partial g_i}{\partial z} - \beta_i^2 g_i \right) \hat{E}^i = 0 \tag{3.31}$$

Using $\hat{E}^i$, which is the solution of the following equation (3.32), and the approximation of (3.33), (3.31) results in (3.34).

$$\frac{\partial^2 \hat{E}^i}{\partial x^2} + \frac{\partial^2 \hat{E}^i}{\partial y^2} + (k^2 n_s^2 - \beta_i^2)\hat{E}^i = 0 \tag{3.32}$$

$$n^2 - n_s^2 = (n_s + q)^2 - n_s^2 \approx 2n_s q \tag{3.33}$$

$$\sum_i \left(\frac{d^2 g_i}{dz^2} - 2j\beta_i \frac{dg_i}{dz} + 2n_s q k^2 g_i \right) \hat{E}^i = 0 \tag{3.34}$$

Equation (3.34) is multiplied by $\hat{H}^{j*}$ and then integrated in the fiber cross-sectional plane S. Using the orthogonality, we obtain

$$\frac{d^2 g_i}{dz^2} - 2j\beta_i \frac{dg_i}{dz} + \frac{k^2}{P} \sum_j g_j \int_s n_s(x,y)\, q(x,y)\, \hat{E}^j \hat{H}^{i*}\, dS = 0 \tag{3.35}$$

where P is the power of each mode. When only one guided mode j is excited and other modes do not exist at $z = 0$ in Figure 3.7, then the solution $g_i(z)$ for (3.35) is interpreted as an amplitude of mode i converted from the initial mode j by the splice. Solution $g_i(z)$ is divided into two parts:

$$g_i(z) = g_i^+(z) + g_i^-(z) \tag{3.36}$$

where $g_i^+(z)$ and $g_i^-(z)$ represent the amplitudes of a wave traveling in the $+z$ and $-z$ directions, respectively. Therefore, c_{ij}, the mode conversion coefficient from mode j to mode i, is expressed by

$$c_{ij} = |g_i^+(d_z)|^2 \tag{3.37}$$

Using (3.35) to (3.37), the following equations are obtained approximately:

$$c_{ij} = f(i,j) \qquad (i \neq j) \tag{3.38}$$

$$c_{jj} = [1 - f(j,j)]^2[1 + f(j,j)] \tag{3.39}$$

where

$$f(i,j) = \frac{k^4 \beta_i^2}{4P^2 \beta_j^2 \omega^2 \mu_0^2} \left| \int_0^d dz \int_s n_s(x,y) q(x,y)\, \hat{E}^j \hat{E}^{i*}\, dS \right|^2 \tag{3.40}$$

Mode conversion from guided mode to radiation modes appears as a loss. Using the mode power P_j for the mode j, the input and output powers P_{in} and P_{out} are expressed by

$$P_{in} = \sum_j P_j \tag{3.41}$$

$$P_{out} = \sum_j \sum_i c_{ij} P_j \qquad (3.42)$$

The symbol Σ_i indicates the summation of all guided modes. Then the loss L caused by the splice is

$$L = -10 \log \left[\frac{\sum_i \sum_j c_{ij} P_j}{\sum_j P_j} \right] \quad \text{(dB)} \qquad (3.43)$$

The baseband frequency response $H(f)$ of spliced fibers is also treated by the mode conversion coefficient c_{ij}. Here we consider the simple case where two fibers are spliced and neither has any mode coupling except for the splicing part. In this case, the baseband frequency response $H(f)$ is expressed by

$$H(f) = \frac{\sum_i \exp(-j2\pi\tau_i f z_2) \left[\sum_j c_{ij} P_j \exp(-j2\pi\tau_j f z_1) \right]}{\sum_j P_j} \qquad (3.44)$$

Where τ_i is the group delay for the mode i, and z_1 and z_2 are the fiber lengths of the spliced two fibers, respectively.

By using the derived equation to evaluate the effect of mode conversion, the following refractive-index derivations $q(x,y,z)$ are assumed for both splice methods.

$$q(x,y,z) = n_V - n_s(x,y) \quad \text{(V-groove splice)} \qquad (3.45)$$

$$q(x,y,z) = n_0[-Q_1 + Q_2\Delta(r/a)^2 + Q_3\Delta(r/a)^4] \sin(\pi z/d_z) \quad \text{(fusion splice)} \qquad (3.46)$$

where n_V, Q_1, Q_2, and Q_3 are constants. n_V represents the refractive index of matching material used in the V-groove splice. In the fusion splice, the derivation of the profile is assumed to be a function of a constant, r^2, and r^4. For simplicity, the mode field function is assumed to be a Laguerre Gaussian function, which is the solution of a wave equation with a simplified assumption of a fiber without considering the cladding effects. For calculations we use an assumed mode power distribution, which approximates a steady-state mode power distribution. In the fusion splice, calculations are made by only varying Q_2, while Q_1 and Q_3 take constant values ($Q_1 = 0$ and $Q_3 = 0.3$). In the V-groove splice, $n_V = n_0$ is used for the calculation. The fiber parameters used for the calculations are $2a = 30$ μm, $\Delta = 1\%$, and $\alpha = 2$.

First we consider the relationship between the splice loss and the mode conversion, shown experimentally in Figure 3.6. The calculated results show that large mode mixing takes place for a splice with large splice loss in both splice methods. It is also demonstrated that the calculated mode mixing of the fusion splice is about 2.8 times larger than that of the V-groove splice for the same splice loss in the calculated splice loss range of 0 to 0.1 dB. Here the mode mixing is defined as the mode conversion power between guided modes divided by the input power. Splices in the experiment in Figure 3.6 have lateral offsets; however, the effect of lateral offsets can be considered the same in both splice methods. Therefore, the calculated results explain the experimental results qualitatively. Next we consider the baseband frequency response $H(f)$ of spliced fibers. Bandwidth expansion E is defined as

$$E = \frac{B - B_0}{B_0} \tag{3.47}$$

where B and B_0 are the bandwidths for spliced fibers with loss and for ideally spliced fibers with no loss and no mode conversion, respectively. For the calculations, $z_1 = z_2 = 1$ km is assumed. For the same type of fibers (i.e., two fibers belong to $\alpha_1 > \alpha_{opt}$ and $\alpha_2 > \alpha_{opt}$, or $\alpha_1 < \alpha_{opt}$ and $\alpha_2 < \alpha_{opt}$), the calculated bandwidth expansion E is very small. For example, E is a few percent for 0.1-dB splice loss. Bandwidth becomes slightly wider for larger splice loss because of mode conversion caused by the splice. When two different kinds of fibers (such as $\alpha_1 > \alpha_{opt}$ and $\alpha_2 < \alpha_{opt}$, or $\alpha_1 < \alpha_{opt}$ and $\alpha_2 > \alpha_{opt}$) are ideally spliced, the spliced fiber has wider bandwidth, which is known as the mode compensation effect, which can be explained by using Figure 3.9(a).

As discussed in Chapter 2, the speed of the lower modes is faster for $\alpha > \alpha_{opt}$ than that of the higher modes, and is slower for a $< \alpha_{opt}$. Therefore, the speed of

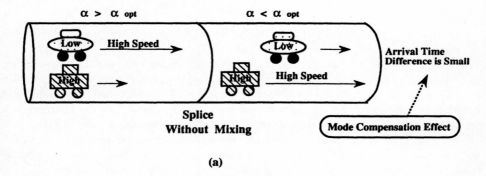

(a)

Figure 3.9 Mode conversion of spliced fibers: (a) ideal splice; (b) real splice.

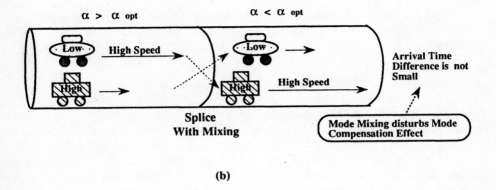

(b)

Figure 3.9 (Continued)

spliced fibers in the case of two different kinds of fibers is averaged, and the speed difference between the lower and higher modes becomes small. This makes the spliced fiber's bandwidth wider (the mode compensation effect). The calculated results of E show that the bandwidth of spliced fibers does not always become wider, but becomes narrower for a large splice loss. Mode mixing with a splice makes the bandwidth wider for the same kinds of fibers. However, the mode mixing disturbs the mode compensation effect and it tends to decrease the bandwidth for different kinds of fibers. These create the complex bandwidth situations of spliced multimode fibers. An example of the complex situation is the measured bandwidth of seven spliced GI multimode fibers shown in Figure 3.10. Due to the mode compensation

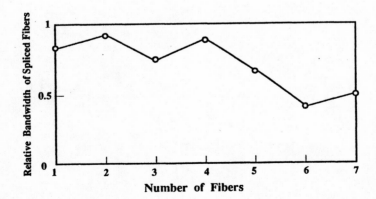

Figure 3.10 Measured bandwidth of spliced fibers.

effect and the mode conversions caused by the fiber itself (such as bends and core-cladding imperfections) and splices, the measured bandwidth shows the complex behavior.

3.2.3 Influence of Fiber Structure on Connection Loss

The fiber structure also influences the connection loss in multimode fibers. Theoretical investigation on a fiber structure anticipates that the excess loss due to an outer layer mainly occurs near excitation points, such as launching and splicing points [11]. This influence on splices has been experimentally investigated [12] and is explained here. The GI multimode fibers used in the experiments are shown in Figure 3.11(a). They have a three-layer structure with various combinations of cladding thickness t and Δ'. The symbol Δ' is defined as

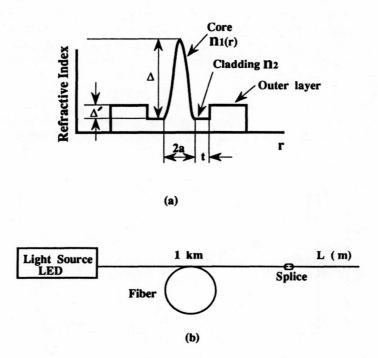

(a)

(b)

Figure 3.11 Structure of optical fiber: (a) fiber structure with a three-layer structure; (b) experimental setup.

$$\Delta' = (n_3 - n_2)/n_2 \qquad (3.48)$$

where n_2 and n_3 are the refractive indexes of a cladding and an outer layer, respectively. Other parameters such as the core diameter $2a = 60$ μm and $\Delta = 1\%$ are same for all fibers used in this experiment. An experimental setup is shown in Figure 3.11(b). The light source is an LED and the input fiber length is approximately 1 km. The splice loss is evaluated by the following procedure:

1. The output power P_0 of a fiber approximately 1 km long is measured.
2. The fiber is cut near the output end.
3. The fibers are spliced with the fusion splice method.
4. Output power P_1 is measured after splicing at the point where the distance from the spliced points is L. In this measurement, P_1 is measured by varying the length L. A splice loss is evaluated by the ratio of P_1 and P_0 as a function of L.

The results are shown in Figure 3.12(a), and various fibers have nearly the same splice losses of about 0.05 dB at the point with $L = 10$ cm after the splicing point. However, the splice losses have different values for different structures for larger L values. Fibers with large Δ' and small t produce a large splice loss. The FFP of the output power is measured for 1-km-long fibers as a measure of the mode power distribution in the fiber and is shown in Figure 3.12(b). The symbol θ represents the propagation angle in air. The measured FFP indicates that there exist fewer higher modes in the fibers with higher Δ' and small t. By comparing Figure 3.12(a,b), it can be seen that fibers with higher Δ' and small t have large splice losses, in spite of the small amount of higher modes. This may be explained by the possibility that the mode conversion from the lower modes to the higher modes and vice versa occurs at the splice point, and that the higher modes attenuate a great deal for fibers with higher Δ' and small t. The measured loss properties due to the lateral offset of various fibers are shown in Figure 3.12(c). These properties can be a measurement to determine an easy-splice fiber or a difficult-splice fiber. The butt joint loss is measured at the point with $L = 20$ m after the joint point. Without lateral offsets, the butt joint loss is set to be 0 dB. It is clear from Figure 3.12(c) that fibers with higher Δ' and small t are difficult-splice fibers. From these experiments, the fiber structure of multimode fibers is found to influence the connection loss.

3.2.4 Loss Properties for Connections in Cascade

Splice or connector loss (generally with passive components used for multimode fibers) depends greatly on the mode power distribution in a multimode fiber, which is affected by the mode conversion caused by connections. When many splices or connectors exist in cascade, the connection loss is affected by the previous connections nearer the light source. Connection loss in cascade is discussed by using the

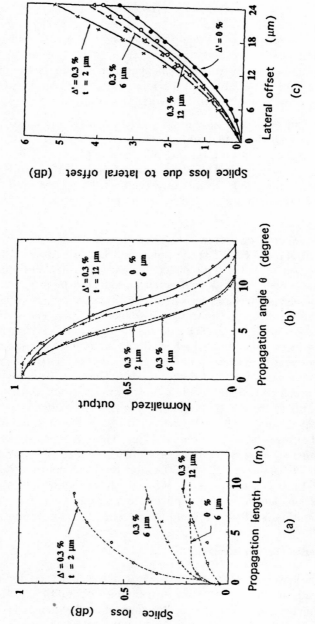

Figure 3.12 Measured splice characteristics for fibers with various structures: (a) length dependence of fusion splice losses; (b) measured FFP; (c) measured splice loss (dB). (After [12].)

power distribution vector and mode conversion matrix [10]. The mode conversion matrix was introduced by (2.71). The mode power distribution vector $[P_i]_0$ satisfying the following equation (3.49) is called the steady-state mode power distribution:

$$[C_{ij}][P_i]_0 = \lambda_e[P_i]_0 \qquad (3.49)$$

where $[P_i]_0$ is the eigenvector (the steady-state mode power distribution) and λ_e is the eigenvalue. When $[C_{ij}]$ includes the mixed effect of mode conversions of a fiber itself and a splice, the mode power distribution for the cascaded splice is expressed by $[C_{ij}]^n[P_i]$, where the same length of fiber and the same splice are assumed for simplicity and the cascaded number is n. If the steady-state mode power distribution for $[C_{ij}]$ is realized, then the following equation holds:

$$[C_{ij}]^n[P_i] = [C_{ij}]^{n-1}\lambda_e[P_i]_0 = \cdots = \lambda_e^n[P_i]_0 \qquad (3.50)$$

In this case, splice loss is not affected by the previous splice and is the same value for all splices in cascade. When the input mode power distribution $[P_i]_{in}$ is very different from $[P_i]_0$, the splice loss is considerably affected by the previous splice until the steady-state mode power distribution is realized.

3.2.5 Modal Noise Caused by Connection

A laser diode has a very narrow linewidth and the emitted light has the property of coherency, which means it can interfere coherently. When the light of a laser diode is launched into a multimode fiber, the excited modes cause the speckle pattern at the fiber endface as a result of the interference among the excited modes [13]. This speckle pattern is very sensitive to fiber state changes, such as external forces (e.g., vibration, pressure) and temperature change. Because the speckle pattern is caused by the coherence of the light source, the variation of the emitting wavelength of a laser diode also makes the pattern change. When there are any spatial filtering effects, such as an imperfect connection, in passive optical components, the transmission loss changes randomly due to the combination of speckle pattern variation and spatial filtering effects [14]. This random loss change causes the noise, called modal noise, in fiber transmission systems. Since optical analog transmission systems require a high signal-to-noise ratio (SNR), modal noise is harmful to analog transmission systems using multimode fibers. Modal noise also occurs in single-mode fiber transmission systems when a single-mode fiber is operated slightly above cutoff [15]. In this case, not only the fundamental mode, but also the next higher modes can transmit in a single-mode fiber. Although the fundamental mode is excited by the light source, mode conversion at the connection point may occur and the interference between the converted higher modes and the fundamental mode creates modal noise. Modal noise in a single-mode fiber transmission can be observed; however, the power penalty from this modal noise is reported to be small when using a sufficiently long fiber or a fiber in a bending state [16]. Since the loss of higher modes is large, higher modes are filtered out in a long or bent fiber.

3.3 REFLECTION PROPERTIES AND INFLUENCE

Reflection properties of optical components are very important for fiber transmission systems, especially in high-bit-rate, analog, or coherent transmission systems. Reflections in these systems degrade system performance. Although an optical isolator can block the reflection to a light source (such as a laser diode), just like an isolator of microwave systems, it is desirable to decrease the reflection of passive optical components. Moreover, an optical isolator cannot prevent the intensity noise that is interferometrically converted from the phase noise of a laser diode by reflections. As was indicated in Figure 1.15, reflection causes crosstalk in bidirectional transmission systems. In this case, an optical isolator also cannot prevent the crosstalk. Reflection at the interface between a fiber and a passive optical component due to the refractive-index difference is roughly estimated by the following well-known equation:

$$R = \left(\frac{n_A - n_B}{n_A + n_B}\right)^2 \tag{3.51}$$

where n_A and n_B are the refractive indexes of two materials. Although this equation is valid for plane waves with perpendicular endface reflections to the light propagation direction, a rough estimation is possible with this equation. Reflection due to splices or connectors occurs because of the refractive-index difference at an interface between a fiber and a spliced part (or connected part).

3.3.1 Reflection Properties of Fusion Splices

The refractive index of the fusion-spliced part is slightly different from that of the ordinary fiber part as discussed in the previous section, and it results in the reflection of light. Reflection of a fusion splice is calculated by (3.51) using Δ as follows:

$$R \approx \left(\frac{n_1 - n_2}{2n_1}\right)^2 = \frac{\Delta^2}{4} \tag{3.52}$$

In this equation, it is assumed that the refractive index of the spliced part is that of a cladding. Therefore, the reflection calculated by (3.52) is overestimated for an ordinary fusion splice. For $\Delta = 1\%$ (multimode fibers) and $\Delta = 0.2\%$ (single-mode fibers), R is calculated as -46 and -60 dB, respectively. Because of the small refractive-index difference, reflection is very small in a single-mode fiber when compared to a multimode fiber with this approximation. The actual reflection may be smaller than these calculated values because the refractive index of the spliced part is similar to that of the original index profile, not that of the cladding in the case of a low-loss fusion splice. If we assume that only lateral offset occurs and that the index profile is not deformed in a fusion-spliced part, we obtain the relation of spliced loss and reflection value with the following calculations [17]. We discuss here the square-low index profile ($\alpha = 2$) multimode fibers.

$$n_A = n_0[1 - \Delta(x^2 + y^2)/a^2]$$ (3.53)

$$n_B = n_0[1 - \Delta((x + d)^2 + y^2)/a^2]$$ (3.54)

Here, we consider that two fibers are laterally offset with d. If we assume the local reflection $R(x,y)$ is obtained by (3.52) to (3.54), the total reflection R_t is expressed by

$$R_t = \frac{\int_s R(x,y)P(x,y)\,dx\,dy}{\int_s P(x,y)\,dx\,dy}$$ (3.55)

where $P(x,y)$ is the mode power distribution. Although the above assumption of $R(x,y)$ is not accurate, the reflection from a fusion splice can be evaluated as a function of splice loss for small splice loss using this simple theory [17]. The calculated results for a fiber with $2a = 50$ μm and $\Delta = 1\%$ are shown in Figure 3.13. The reflection from a fusion splice with 0.1-dB loss for the steady state is estimated to be about -68 dB. When compared to -46 dB for $\Delta = 1\%$ from (3.52), it is under about 20 dB. The measured reflection of fusion-spliced GI fibers is shown in Figure 3.14 [17]. The reflection is measured by an OTDR (explained in Chapter 4). Since the reflection value for a single splice is very small, the reflection of 12 cascaded splices with a total loss of 5 dB is used in the experiment. The experimental results are represented by a circle in the figure, and the numeral 12 near the circle indicates the number of closely located fusion splices in cascade. It is measured to be about -50 dB and agrees well with the theoretical curve predicted by (3.55). Therefore, the reflection of a fusion-spliced GI fiber with 0.1-dB loss is estimated to be -68 dB. As for a single-mode fiber with a splice loss under 0.1 dB, the reflection is estimated to be well under -70 dB.

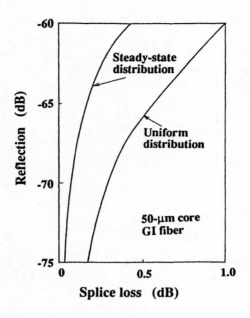

Figure 3.13 Calculated relationship between reflection and splice loss in fusion splices. (After [17].)

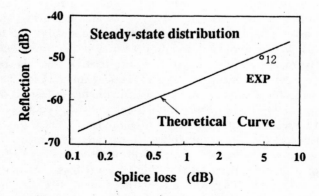

Figure 3.14 Measured relationship between reflection and splice loss in fusion splices. (After [17].)

3.3.2 Reflection Properties of Connectors

We discuss here the reflection of connectors, mechanical splices, and adhesive-bonded splices. In this section, physical or optical contacts between fibers are not considered; connections with physical or optical contacts are discussed in the next section. Connections with a gap between fibers are discussed here. The gap between fibers is filled with air or matching materials in ordinary connectors, mechanical splices, and adhesive-bonded splices. The air-gap connector, which uses no index-matching material and whose gap is not small, has a reflection with a value of about -15 dB. This value is calculated by using (3.51) with $n_A = 1$ and $n_B = 1.46$. Usually, index-matching material is applied to lower the loss and reflection. The reflection of the connection using index materials can be estimated by (3.51) or (3.55). When using (3.55), $R(x,y)$ must be obtained by (3.51), (3.53), and $n_B = n_V$ (constant), where n_V is the refractive index of matching materials. Therefore, it is anticipated that the reflection depends on the refractive index of the matching material and does not depend on the lateral offset (i.e., the connection loss). This situation is different from that of a fusion splice. Using the experimental setup shown in Figure 3.15(a) and an OTDR, the relationship between the reflection and the lateral offset is measured. Two types of matching material, $n_V = 1.429$ and $n_V = 1.469$, are used for the measurement. The gap z is less than a few microns, but the physical contact (i.e., the fiber endfaces have physical contact) is not realized. The results are shown in Figure 3.15(b). These results confirm that the reflection from connection methods using index-matching materials depends on the matching material's index, not on the lateral offset. The measured fluctuations of reflection may be the interference effect caused by the multiple reflections between the fiber endfaces, which cause not only the reflection fluctuation, but also the connection loss variation when reconnected. Multiple reflections between fiber endfaces, shown in Figure 3.16, are treated like a Fabry-Perot interferometer. Two fiber endfaces make up the Fabry-Perot resonator. Light from a fiber on the left incidents into a gap and passes through the gap. Some portion of light reflects at the endface of a fiber on the right and passes through a gap and again reflects, and so on. These multiple reflections are expressed as follows: the first transmitted light is $E_i t_1 t_2 e^{-\gamma S}$, the second transmitted light is $E_i t_1 r_1 r_2 t_2 e^{-3\gamma S}$, and so on.

The notations E_i, t_1, t_2, r_1, r_2, γ, and S are the electric field of incident light, the transmittance coefficient at the left fiber endface, the transmittance coefficient at the right fiber endface, the reflection coefficient at the left fiber endface, the reflection coefficient at the right fiber endface, the complex propagation constant, and the gap length, respectively. The field of output light from a right fiber endface is the sum of these transmitted lights and is

$$E_t = E_i t_1 t_2 e^{-\gamma S}[1 + r_1 r_2 e^{-2\gamma S} + (r_1 r_2 e^{-2\gamma S})^2 + \ldots]$$

$$= \frac{t_1 t_2 e^{-\gamma S}}{1 - r_1 r_2 e^{-2\gamma S}} E_i \tag{3.56}$$

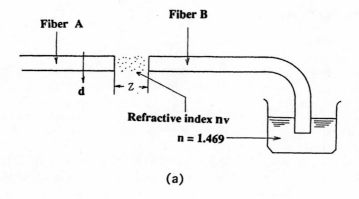

(a)

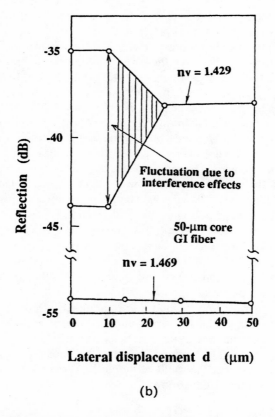

(b)

Figure 3.15 Measured relationship between reflection and lateral offset in splices using matching material: (a) experimental setup; (b) measured reflection and lateral offset. (After [17].)

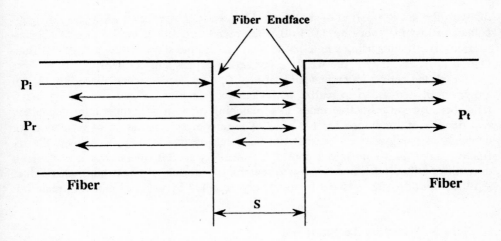

Figure 3.16 Multiple reflections between fiber endfaces.

and the ratio of transmitting power to input power is

$$\frac{P_t}{P_i} = \frac{|E_t|^2}{|E_i|^2} = \frac{|t_1 t_2\, e^{-\gamma S}|^2}{|1 - r_1 r_2\, e^{-2\gamma S}|^2} \tag{3.57}$$

The propagation constant γ is

$$\gamma = \alpha/2 + j\beta \approx j2\pi n/\lambda \tag{3.58}$$

where α and β are the optical loss in gap and the phase constant, respectively.

The phase constant β is $2\pi n/\lambda$, where n is the refractive index of the gap. We assume $\alpha = 0$ (no loss). Using (3.58), the calculated result of (3.57) is

$$\frac{P_t}{P_i} = \frac{(1 - R)^2}{(1 - R)^2 + 4R\, \sin^2(2\pi n S/\lambda)} \tag{3.59}$$

where the following relationships between reflection coefficients for amplitude and reflection coefficients for power are used:

$$r_1^2 = r_2^2 = R \tag{3.60}$$

The transmitting power P_t and the reflected power P_r vary with the gap period of $\lambda/2n$. For example, this simple theory predicts that connection loss varies by ± 0.3

dB with the air gap period of $\lambda/2$ ($n = 1$). As for connectors with a large air gap, connector loss may vary by ±0.3 dB when reconnected. The connection and reconnection experiments using single-mode fiber connectors show ±0.2- to 0.25-dB fluctuation [18], and they agree with the prediction of this simple theory.

With the perfect index matching of $R = 0$, no reflection is anticipated by (3.51). However, it is difficult to realize perfect index matching by using matching materials, because the refractive index in a matching material is temperature-dependent and fiber facet conditions are not ideal. The temperature variation of reflection for connections using a matching material has been measured [17]. Moreover, the refractive index in a core is not constant, especially for GI fibers, and it is difficult to realize perfect index matching with matching materials of a constant index value. A reflection of under -45 dB is practically obtained by using a matching material.

3.3.3 Low-Reflection Techniques

Generally, an optical connection has a relatively large reflection value, except for that made by the fusion splice method. Low-reflection techniques for optical connectors, mechanical splices, and adhesive-bonded splices have been developed.

To achieve low reflection, the following methods are considered:

1. Suppression of reflection by index matching with matching materials;
2. Suppression of reflection by index matching with a physical contact (no gap; i.e., $S = 0$);
3. Filtering out the reflected light with the obliquely formed fiber endface.

As discussed in the previous section, the low-reflection technique of method 1 also serves to suppress the connection loss fluctuation in the case of connection and reconnection. Matching gel and matching oil are generally used for matching materials. Methods 2 and 3 are explained by using the case of connectors shown in Figure 3.17, which shows low-reflection techniques for optical connectors without index-matching material. As explained previously, the contact state in which two plugs are mated without a gap is called physical contact (PC). This is realized by a sophisticated polishing technique and realizes both low loss and low reflection because of there being no index difference between the two fibers. A measured reflection of about 28 dB using this technique was reported for single-mode fiber PC connectors [18], which indicates about 13 dB improvement when compared with air gap connectors. To reduce the reflection further, the refractive-index change of the fiber surface through polishing must be considered. The estimated refractive index of the fiber surface polished with diamond powder ranges from 1.5 to 1.6, and it depends on the polishing particle diameters [19]. It is known that a small refractive-index change of a silica glass surface is realized when it is polished with a cerium grain having low hardness [20]. The use of this advanced polishing technique allows

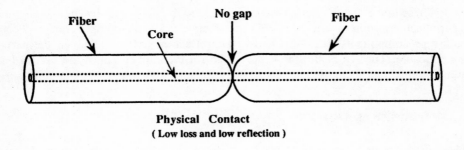

Physical Contact
(Low loss and low reflection)

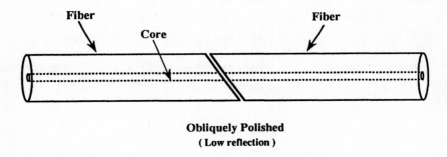

Obliquely Polished
(Low reflection)

Figure 3.17 Low-loss and low-reflection techniques for optical connectors.

a return loss of −40 to −50 dB to be obtained for the physical contact connectors. The obliquely polished connectors have been proposed as an alternative technology, and the polished angles θ for single-mode and multimode fibers must be larger than about 5 and 9 deg, respectively, to obtain a −40-dB reflection value [21]. Although the obliquely polished connector generally has larger loss than ordinary connectors, it has the lowest reflection among the proposed low-reflection techniques. A reflection of about −60 dB for single-mode fiber connectors is realized. It is possible to use a combination of the oblique polishing and matching material technologies and a combination of the oblique polishing and physical contact technologies. Both physical contact and accurate oblique polishing are difficult to apply for array connectors (multifiber connectors) because of the wide surface area for polishing. Recently, the combination of oblique polishing and physical contact technologies has been developed, and it realizes a single-mode four-fiber array connector with low insertion loss (0.16 dB) and low reflection loss (−59 dB reflection) [22].

These low-reflection technologies developed mainly for optical connectors may be applicable to other passive optical components. For some components (especially bulk-type components), the well-known antireflection (AR) coating on the surface is applicable. The AR coating is widely used for the application of the optical lens, and it is made by using dielectric multilayers.

3.3.4 Influence of Reflection

Many fiber transmission systems use laser diodes, and even a small amount of optical feedback or light injection changes the lasing characteristics of laser diodes. The characteristics influenced by reflections are:

1. Increase of output intensity noise [23–26];
2. Enhancement of harmonic modulation distortions [27,28];
3. Variation of linewidth [29–32].

The influence of reflections is generally larger for single-mode laser diodes than for multimode laser diodes.

It was reported that a laser diode coupled with a fiber has periodic sharp peaks in the noise spectra, which are the fundamental peak frequency f_0 and its overtone mf_0 [23]. The peak frequency f_0 is

$$f_0 = c/(2nZ_f) \qquad (3.61)$$

where c, n, and Z_f are light velocity in a vacuum, the fiber refractive index (n is about 1.46 for silica fiber), and the fiber length, respectively. These peaks are caused by the reflection of the far-end fiber facet, and they cause the intensity noises. For a 1-km fiber, $f_0 = 100$ kHz. This type of reflection causes submodes in a single-mode laser spectrum, which is originally a smooth Lorentzian shape [24]. The frequency space between submodes corresponds to the round-trip time of the optical external resonator, which is formed by a laser facet and a fiber endface. The noise peaks may also occur because of the reflection of passive optical components attached to a fiber. Reflection near a laser induces the low-frequency intensity noise [25,26]. This noise increases along with the increase of reflection [25].

Laser output power changes because of the reflection near a laser, and the current-light curve (I-L curve) tends to have nonlinear properties. This causes the higher order harmonic distortions of the modulated laser light [27,28]. The distortions degrade system performance especially in analog transmission systems, where accurate waveform transmission is required.

Reflection causes lasing spectrum changes, such as changes in lasing wavelength and linewidth. Mechanically tunable laser diodes in external cavity laser diodes are realized based on these effects caused by the reflection. The general structure of mechanically tunable external cavity laser diodes uses the ordinary laser chip and an external cavity which has wavelength selectivity. One example of the selective device is a grating. The laser diode chip is usually AR coated on the external cavity side to make use of the reflected light efficiently. The laser linewidth is also changed by reflection and it is broadened or narrowed. Very narrow linewidth is realized in external cavity laser diodes. The reflection properties of laser diodes depend on the laser diode's linewidth enhancement factor [31]. Although these lasing spectrum changes are useful for external cavity laser diodes, they degrade the transmission performance, especially in coherent transmission systems.

The influences mentioned above are laser diode property changes caused by reflections. There is reflection-induced transmission performance degradation not only by laser diode property change, but also by other factors in some systems such as bidirectional and coherent transmission systems. Reflection causes crosstalk in bidirectional transmission systems. Both the signal and reflected light enter a photodetective device, such as a photodiode. The crosstalk caused by the reflected light degrades the minimum detection level. The phenomenon of reflection forms an equivalent interferometer in the transmission systems, and the intensity noise is interferometrically converted from the phase noise of a laser diode by the formed equivalent interferometer [33]. Although an optical isolator is useful for suppressing the reflection effect in many cases, an isolator cannot prevent the crosstalk in bidirectional transmission systems and interferometrically converted intensity noise.

The required suppression of reflection value depends on the transmission systems. Here, some examples of the required reflection values are explained. In the case of bidirectional transmission systems, the extinction ratio E_x is degraded due to the reflected light. The extinction ratio E_x is defined by (3.63). The required input light to a detector increases ΔP (the minimum detectable light increases ΔP) in the case of a finite E_x value when compared to the ideal infinite E_x value. ΔP is expressed as [34]

$$\Delta P = 10 \log \left(\frac{E_x + 1}{E_x - 1} \right) \qquad \text{(dB)} \qquad (3.62)$$

$$E_x = P_1/P_0 \qquad (3.63)$$

where P_1 and P_0 are the signal and crosstalk lights in this case. When $\Delta P < 0.1$ dB, $E_x = 85$, and this means the reflection must be lower than about -20 dB. When the number of reflections is N and its distance is Z_i ($i = 1, 2, \ldots, N$), then the total reflection R_t is

$$R_t = \sum_i R_i \exp(-2\alpha Z_i) \qquad (3.64)$$

where R_i and α are the reflection at point i and the fiber loss, respectively. This equation assumes that the reflections are added incoherently. In a simple case of αZ_i = 0 and $R_i = R$, $R_t = NR$. For $N = 10$, the required reflection is under -30 dB in the case of $\Delta P < 0.1$ dB. These equations assume out-band interference, which means that the frequencies or wavelengths upstream and downstream are different. In the case of intersymbol interference, the required reflection may be more severe. In the high-bit-rate transmission of IM/DD, the relative intensity noise (RIN) degrades the receiver sensitivity. The RIN is defined as the ratio of the output light intensity fluctuation to the averaged light intensity. The required RIN for 4 Gbps is reported to be -140 dB/Hz for a power penalty of 0.1 dB [35]. With the calculation of RIN, a reflection of under -50 dB is required for a laser diode with 1 mW of output power in the case of the linewidth enhancement factor $\alpha = 6$ [36]. Using R_t = NR, R must be under -60 dB for $N = 10$. When the output power of a laser diode is higher, the value is moderated. For example, R is moderated to be under -30 dB $(N = 1)$ for a 10-mW power laser with $\alpha = 6$ in the case of a 0.1-dB power penalty using the calculated RIN value in [36]. In the case of coherent transmission systems, the requirement for linewidth is severer than that in IM/DD systems. Although it depends on transmission methods such as homodyne and heterodyne detection [37], the required reflection level may range from -60 to -80 dB. It is reported that an isolator with 50 dB is required when using a DFB-LD [32].

REFERENCES

[1] Marcuse, D., "Loss Analysis of Single-Mode Fiber Splices," Bell Syst. Tech. J., Vol. 56, 1977, p. 703.

[2] Gloge, D., "Offset and Tilt Loss in Optical Fiber Splices," Bell Syst. Tech. J., Vol. 55, 1976, p. 905.

[3] Thiel, F. L. and R. M. Hawk, "Optical Waveguide Cable Connection," Appl. Opt., Vol. 15, 1976, p. 2785.

[4] Neumann, E. G. and W. Weidhaas, "Loss Due to Radial Offsets in Dielectric Optical Waveguides With Arbitrary Index Profiles," A.E.Ü., Vol. 30, 1976, p. 448.

[5] Miller, C. M. and S. C. Mettler, "A Loss Model for Parabolic-Profile Fiber Splices," Bell Syst. Tech. J., Vol. 57, 1978, p. 3167.

[6] Seikai, S. and N. Uchida, "Structural Parameter Specifications of a Graded-Index Fiber on the Basis of Splice Loss," IECE of Japan, Vol. E-65, 1982, p. 485.

[7] Kashima, N. and N. Uchida, "Relation Between Splice Loss and Mode Conversion in a Graded-Index Optical Fiber," Electron. Lett., Vol. 15, 1979, p. 336.

[8] Kashima, N., "Splice Loss and Mode Conversion in a Multimode Fiber," IECE, Technical Report OQE 80–18, 1980, p. 7 (in Japanese).

[9] Kashima, N., "Splice Loss and Mode Conversion in a Multimode Fiber," Appl. Opt., Vol. 19, 1980, p. 2597.

[10] Kashima, N., "Transmission Characteristics of Splices in Graded Index Multimode Fibers," Appl. Opt., Vol. 20, 1981, p. 3859.

[11] Kashima, N. and N. Uchida, "Transmission Characteristics of Graded-Index Optical Fibers With a Lossy Outer Layer," Appl. Opt., Vol. 17, 1978, p. 1199.

[12] Kashima, N., N. Uchida, N. Susa, and S. Seikai, "The Influence of a Fiber Structure on Optical Loss in a Graded-Index Fiber," Electron. Lett., Vol. 14, 1978, p. 78.

[13] Epworth, R. E., "The Phenomenon of Modal Noise in Analog and Digital Optical Fiber Systems," European Conf. on Optical Communications (ECOC'78), 1978, p. 492.

[14] Petermann, K., "Nonlinear Distortion and Noise in Optical Communication Systems Due to Fiber Connectors," IEEE J. Quantum Electron., Vol. QE-16, 1980, p. 761.

[15] Heckmann, S., "Modal Noise in Single Mode Fibers Operated Slightly Above Cutoff," Electron. Lett., Vol. 17, 1981, p. 499.

[16] Cheung, N. K., A. Tomita, and P. F. Glodis, "Observation of Modal Noise in Single-Mode-Fiber Transmission Systems," Electron. Lett., Vol. 21, 1985, p. 5.

[17] Kashima, N. and I. Sankawa, "Reflection Properties of Splices in Graded-Index Optical Fibers," Appl. Opt., Vol. 22, 1983, p. 3820.

[18] Suzuki, N., M. Saruwatari, and M. Okuyama, "Low Insertion- and High Return-Loss Optical Connectors With Spherically Convex-Polished End," Electron. Lett., Vol. 22, 1986, p. 110.

[19] Sankawa, I., T. Satake, N. Kashima, and S. Nagasawa, "Fresnel Reflection Reducing Methods for Optical-Fiber Connector With Index Matching Materials," IEICE of Japan, J67-B, 1984, p. 1423. (In Japanese.) English translation is in Electronics and Communications in Japan, Part 1, Vol. 69, Scripta Technica, Inc.,1986, p. 94.

[20] Yokota, H., "Polarization Analysis of Solid and Liquid Surface Layers," Oyo Butsuri (Applied Physics), Vol. 40, 1971, p. 1024 (in Japanese).

[21] Suzuki, N. and O. Nagano, "Low Insertion Loss and High Return-Loss Optical Connectors for Use in Analog Video Transmission," Int. Conf. Integrated Opt. Fiber Commun. (IOOC'83), 1983, p. 30A3–5.

[22] Nagasawa, S., Y. Yokoyama, F. Ashiya, and T. Satake, "A High-Performance Single-Mode Multifiber Connector Using Oblique and Direct Endface Contact Between Multiple Fibers Arranged in a Plastic Ferrule," IEEE Photon. Technol. Lett., Vol. 3, 1991, p. 937.

[23] Ikushima, I. and M. Maeda, "Self-Coupled Phenomena of Semiconductor Lasers Caused by an Optical Fiber," IEEE J. Quantum Electron., Vol. QE-14, 1978, p. 331.

[24] Ikushima, I. and M. Maeda, "Lasing Spectra of Semiconductor Lasers Coupled to an Optical Fiber," IEEE J. Quantum Electron., Vol. QE-15, 1979, p. 844.

[25] Hirota, O. and Y. Suematu, "Noise Properties of Injection Lasers Due to Reflected Waves," IEEE J. Quantum Electron., Vol. QE-15, 1979, p. 142.

[26] Hirota, O., Y. Suematu and K. Kwok, "Properties of Intensity Noises of Laser Diodes Due to Reflected Waves From Single-Mode Optical Fibers and Its Reduction" IEEE J. Quantum Electron., Vol. QE-17, 1981, p. 1014.

[27] Lang R. and K. Lobayashi, "External Optical Feedback Effects on Semiconductor Injection Laser Properties," IEEE J. Quantum Electron., Vol. QE-17, 1980, p. 347.

[28] Kikushima, K., O. Hirota, M. Shindo, V. Stoykov, and Y. Suematsu, "Properties of Harmonic Distortion of Laser Diodes With Reflected Waves," J. Opt. Commun., Vol. 3, 1982, p. 129.

[29] Miles, R. O., A. Dandridge, A. B. Tveten, H. F. Taylor, and T. G. Giallorenzi, "Feedback-Induced Line Broadening in CW Channel-Substrate Planar Laser Diodes," Appl. Phys. Lett., Vol. 37, 1980, p. 990.

[30] Kikuchi, K. and T. Okoshi, "Simple Formula Giving Spectrum-Narrowing Ratio of Semiconductor Laser Output Obtained by Optical Feedback," Electron. Lett., Vol. 18, 1982, p. 10.

[31] Patzak, E., H. Olessen, A. Sugimura, S. Saito, and T. Mukai, "Spectral Linewidth Reduction in Semiconductor Lasers by an External Cavity With Weak Optical Feedback," Electron. Lett., Vol. 19, 1983, p. 938.

[32] Tkach, R. W. and A. R. Chraplyvy, "Linewidth Broadening and Mode Splitting Due to Weak Feedback in Single-Frequency 1.5 μm Lasers," Electron. Lett., Vol. 21, 1985, p. 1081.

[33] Choy, M. M., J. L. Gimlett, R. Welter, L. G. Kazovsky, and N. K. Cheung, "Interferometric

Conversion of Laser Phase Noise to Intensity Noise by Single-Mode Fiber-Optic Components," Electron. Lett., Vol. 23, 1987, p. 1151.

[34] Kashima, N., "Optical Transmission for the Subscriber Loop," Norwood, MA: Artech House, 1993.

[35] Shikada, M., S. Takano, S. Fujita, I. Mito, and K. Minemure, "Evaluation of Power Penalties Caused by Feedback Noise of Distributed Feedback Laser Diodes," IEEE J. Lightwave Technol., Vol. 6, 1988, p. 655.

[36] Schunk, N. and K. Petermann, "Numerical Analysis of the Feedback Regimes for a Single-Mode Semiconductor Laser With External Feedback," IEEE J. Quantum Electron., Vol. 24, 1988, p. 1242.

[37] Linke, R. A. and A. H. Gnauck, "High-Capacity Coherent Lightwave Systems," IEEE J. Lightwave Technol., Vol. 6, 1988, p. 1750.

Chapter 4

Measurements for Passive Optical Components

Explained in this chapter are the measuring methods for several characteristics of passive optical components. These are the measuring methods for insertion loss, baseband frequency response or chromatic dispersion, reflection, fiber endface quality, crosstalk, and fault location and diagnosis. Since fiber parameters such as MFDs are important for evaluating the characteristics of passive optical components, their measurements are also explained briefly. In most cases, the baseband frequency response or the chromatic dispersion is not so important for evaluating passive optical components themselves. However, it is very important for fiber transmission, so the baseband frequency response or the chromatic dispersion must be evaluated for total transmission lines, including passive optical components. And it may also be important for the fiber-type components with a sufficient fiber length. As discussed in the previous chapter, transmission properties depend on the mode power distribution in the case of multimode fibers. The mode power distribution must be taken into consideration for measuring both loss and baseband frequency response in multimode fibers. In principle, it is simple to measure loss and baseband frequency response in single-mode fibers. In practice, it is very difficult to measure baseband frequency response in single-mode fibers because of the wide bandwidth. Instead of measuring the baseband frequency response, the chromatic dispersion is measured to evaluate the frequency response. The fiber endface quality measurement is important for a connector, and it is also important for evaluating a performance of fiber-cutting tools. To obtain low-loss splices, high-quality fiber endfaces made by a cutting tool are required. Crosstalk degrades the transmission system performance, and it possibly occurs in passive optical components such as a directional coupler and a filter used for a WDM transmission system. Fault location and diagnosis of passive optical components are important for detecting the broken points or the high-loss points. High resolution in distance or position is required for the measurements used for passive optical components.

4.1 FIBER PARAMETER MEASUREMENTS

Fiber parameters greatly influence the characteristics of passive optical components. For example, splice loss depends on the MFD difference of two mated single-mode fibers, as discussed in the previous chapter. Before evaluating passive optical components, we must know the parameters of the fibers that are used or attached to the components. The fiber parameters, which will influence the performance of passive optical components, are listed in Table 4.1. Usually, the fiber parameters are measured in a factory and the measurement data are attached to the fiber. If the data are missing or if we want to measure the data precisely, we must measure them ourselves.

4.1.1 Single-Mode Fiber Parameter Measurements

MFD 2ω of single-mode fibers is a suitable parameter to evaluate the component's performance (such as connection loss), rather than core diameter 2a, relative refractive-index difference Δ, or refractive-index profile, as discussed in Chapter 2. Many MFD measurement methods have been proposed so far. Some of them are:

1. Far-field pattern method;
2. Near-field pattern method;
3. Transversal offset method;
4. Variable circular aperture launch method;
5. Far-field aperture methods, including knife-edge method;
6. Mask method.

Table 4.1
Important Fiber Parameters

Single-Mode Fibers	GI Multimode Fibers
MFD	Core diameter 2a Refractive-index difference Δ Index profile (α parameter)
	Core noncircularity
Core eccentricity	
Outer diameter	
Outer surface noncircularity	

The FFP and NFP were shown in Figure 2.2 and explained in Chapter 2. These are the measurements of far- and near-field intensity of the fiber mode. The transversal offset method uses the relationship of connection loss and transversal offset (lateral offset) [1]. It makes use of (3.16) to determine the mode field radius ω. The variable circular aperture launch method uses the circular aperture by varying its circle, and the transmitted light through the aperture is used to launch a fiber. The mode field radius ω is determined by measuring the launched power [2]. The far-field aperture method uses several types of apertures, such as circular, slit, and knife-edge apertures. The transmitted light through the aperture is measured to determine ω. The knife-edge method, which uses a knife-edge aperture, is well known to measure the spot size of a laser such as a He-Ne gas laser. The mask method uses a special mask whose transparency (transmittivity) is proportional to the square of the distance from the axis [3]. The near field of the fiber is magnified and this mask is located after the magnified near field. The transmitted light through the mask is measured to determine ω. Although a single parameter ω is not sufficient to determine the characteristics of a mode field except for a Gaussian field distribution, ω is convenient for describing a single-mode fiber.

The accuracy of the diameter dimension is very important for fiber-alignment methods using the outer diameter. It is also important for many connectors, where a hole is used to fix a fiber in a connector ferrule. For a fiber with the outer diameter larger than the hole diameter, these types of connectors cannot be fabricated. For a fiber with a very small outer diameter, the dimensional difference between a fiber diameter and a hole diameter influences connection loss. Core eccentricity is defined as the distance between the two centers of a core and a reference surface (an outer surface layer), divided by the core diameter. The outer diameter and core eccentricity are measured by magnifying the fiber endface. A microscope with a TV camera (such as a vidicon camera) is often used for the magnifying method, and the dimensions are measured digitally by using cursors (Fig. 4.1). These measurements are usually automated with a computer in a fiber manufacturing factory. Noncontact measurement of the outer diameter by a scattering method with a laser beam is used to control the fiber diameter in a fiber drawing process from a mother rod. Due to the progress of fiber manufacturing techniques, the dimensional fiber parameters such as outer diameter and nonconcentricity are well controlled today. Therefore, we can splice many fibers in single-mode fiber ribbons simultaneously with low splice losses using the mass-fusion splice method, which relies on the accuracy of fiber outer diameters.

4.1.2 Multimode Fiber Parameter Measurements

The relative refractive-index difference Δ and refractive-index profile $n(r)$ (or α-parameter) are very important, because these parameters determine the bandwidth

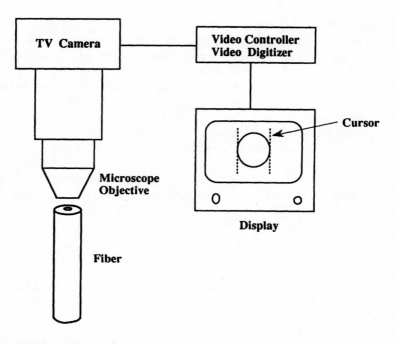

Figure 4.1 Fiber dimensional measurement.

of multimode fibers. Many index profile measurement methods have been proposed and some of them are:

1. Interferometric method using a slab sample;
2. Near-field method;
3. Reflection method;
4. Refraction method;
5. Transverse interferometric method;
6. Scattering method.

The interferometric method using a slab sample uses the interference microscope [4–6]. A fiber sample forming a thin slab is placed in one branch of an optical path, and the homogeneous reference slab is placed in another branch of an optical path, as shown in Figure 4.2. Two beams traveling on two optical paths interfere and cause interference fringes, which are the result of the phase shift difference between a fiber slab and a reference slab. Since the reference slab is homogeneous, the phase shift or fringe shift S corresponds to the refractive-index profile $n(r)$. Fringe interval D in the cladding corresponds to a phase difference of 2π. Shift S in Figure

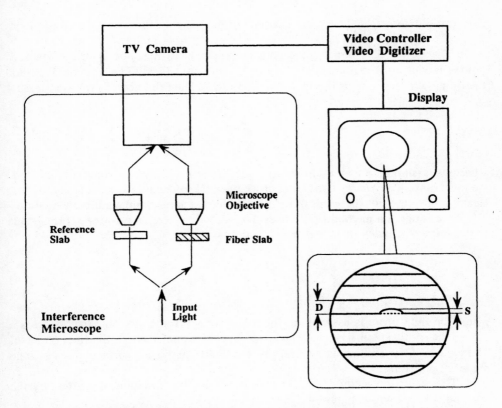

Figure 4.2 Refractive-index profile measurement (interferometric method using a slab sample).

4.2 corresponds to the phase shift of Δnkd, where d is the sample (or reference) slab thickness, $\Delta n = [n(r) - n(a)]$ and $k = 2\pi/\lambda$. The ratio S/D is

$$\frac{S}{D} = \frac{[n(r) - n(a)]kd}{2\pi} = \frac{[n(r) - n(a)]d}{\lambda} \tag{4.1}$$

and the profile $n(r) - n(a)$ is given by

$$n(r) - n(a) = \frac{S\lambda}{Dd} \tag{4.2}$$

Since S is obtained as a function of r, the refractive-index profile is obtained from (4.2). The cladding refractive index $n(a)$ is known from the index measurement, and

Δ is also obtained from (4.2). An example of the fringe picture was shown in Figure 3.8.

The power distribution in a core corresponds to the index profile when all guided modes are equally excited (uniform excitation) [7,8]. The near-field method uses this fact, and the profile is obtained by measuring the power $p(r)$; that is,

$$n^2(r) - n^2(a) = [n^2(0) - n^2(a)] \frac{p(r)}{p(0)} \tag{4.3}$$

where we assume the maximum index is located at $r = 0$. When using this method, we must make leaky mode correction or we must take care not to excite leaky modes. Leaky modes are not the guiding modes and have a large propagation attenuation.

The reflection method [9,10] uses the fact that reflection power P_r corresponds to the refractive index of the fiber endface; that is,

$$P_r = \left(\frac{n(r) - 1}{n(r) + 1}\right)^2 P_{in} \tag{4.4}$$

where the light beam is assumed to input and reflect in the air. This method uses a laser light with a small spot size. The laser light is scanned across the fiber endface. The cleaved fiber endface is used in this method to avoid the surface index change made by polishing.

The refraction method [11,12] uses the following equation, described by the ray optics as shown in Figure 4.3.

$$n^2(r) - n_2^2 = n_2^2 [\sin^2 \theta_1 - \sin^2 \theta_2] \tag{4.5}$$

where $n_2 = n(a)$ and index matching with refractive index n_2 is used. The opaque screen shown in Figure 4.3 is used to suppress the leaky rays. Angles θ_1 and θ_2 are shown in Figure 4.3(b). From (4.5), $n(r)$ is measured.

The transverse interferometric method [13–15] uses interferometry like the interferometric method using a slab sample (method 1), and this method does not require the sample preparation, because the axis of the sample fiber is only placed in the matching oil transversely to the light used for the interferometry. The principle of this method is similar to the method 1, and the phase difference caused by the refractive-index profile $n(r)$ is measured as a fringe shift.

The scattering method [16–18] uses the forward scattering pattern of a fiber. Since the refractive-index profile is expressed by the Hankel transform of the measured scattered field, the refractive index is obtained by the calculation from the measured pattern.

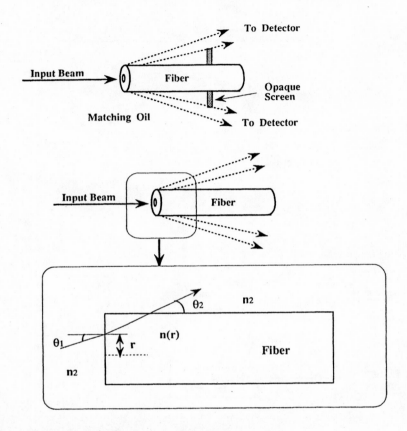

Figure 4.3 Refractive-index profile measurement (refraction method).

As for the geometrical measurements, such as outer diameter measurement, the same measurement methods explained for a single-mode fiber can be used for a multimode fiber. The dimensions of some fiber parameters such as core and cladding diameters can be measured by the near-field method, which is used for the index profile measurement.

4.2 LOSS MEASUREMENT

4.2.1 Cutback and Insertion Methods

The cutback method is an accurate loss measurement method used for a fiber and a passive optical component attached to a fiber. Although this method is very accurate,

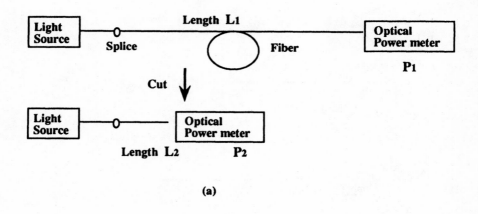

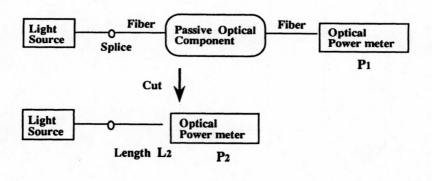

(a)

(b)

Figure 4.4 Cutback method: (a) fiber; (b) passive optical component.

the drawback is its destructive nature; that is, the fiber is cut and becomes short. The method is shown in Figure 4.4(a,b), respectively, for a fiber and a passive optical component. In the case of a fiber with length L_1 in kilometers, the attenuation α is expressed with the powers P_1 and P_2 measured usually by an optical power meter.

$$\alpha = \frac{10}{L_1 - L_2} \log_{10} \left[\frac{P_2}{P_1}\right] \quad \text{(dB/km)} \tag{4.6}$$

where L_2 is the remaining short fiber length after the cut. When the power is measured in decibels referred to 1 mW (dBm), the attenuation is simply $\alpha = [P_2 - P_1]/[L_1 - L_2]$. For a component, insertion loss α is

$$\alpha = 10 \log_{10} \left[\frac{P_2}{P_1}\right] \quad \text{(dB)} \tag{4.7}$$

Since the loss of a fiber with a short length is very small, it is neglected. We must take care not to change the input condition (launching condition) before and after the fiber cut. The reason for the fiber cut is that the coupling power from a light source into a fiber or a component is not constant and the input power must be measured using the short cut fiber. Although the cutback method is very accurate, there are some measurement errors. The main origins of errors are a change of input condition, the effect of the cladding mode, a badly cut fiber endface, and the non-uniform photo-sensitivity of the photodetector surface used in a power meter. The cladding modes attenuate strongly and exist near the input. To suppress the error of the cladding mode effect, the cladding modes must be removed by using a matching oil or some other means. The accuracy of the cutback method is generally ±0.03 dB. When further accuracy is required, we obtain the average value of the losses, which are measured repeatedly.

An alternative method for a loss measurement is the insertion method. The principle of this method is the same as that of the cutback method. The method is shown in Figure 4.5(a,b). The attenuation or loss α is obtained by (4.6) or (4.7). An optical cord is commonly used for measuring the input power and its length is short. An example length is 2 m long. An optical cord is composed of a coated fiber and a sheath with reinforcement materials, and it is designed for easy handling. The input power P_2 is measured by using an optical cord in order to eliminate the influence of connector loss. However, generally the losses of connectors attached to a fiber (or an optical component) and to an optical cord are different. Therefore, the insertion method is not as accurate when compared to the cutback method. It is convenient to use an optical cord for handling; however, it is possible to use a coated

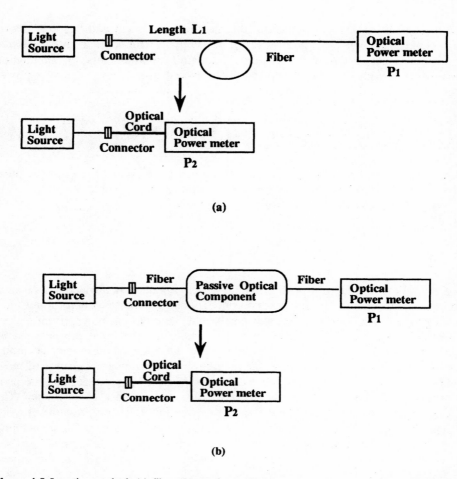

Figure 4.5 Insertion method: (a) fiber; (b) passive optical component.

fiber instead of an optical cord. When connector loss is small, or the accuracy of measurement is acceptable within connector loss, P_2 can be directly measured by an optical power meter without using the optical cord. The application of the insertion method is limited to fibers with an optical connector or to passive optical components with an optical connector. Usually optical connectors are attached to passive optical components used for fiber transmission systems; therefore, this limitation does not practically restrict the application of this method.

A modified insertion method is possible by using an optical coupler as shown in Figure 4.6. When the coupling ratio K is known, the attenuation or loss is obtained by measuring P_1 and P_2. For a component, insertion loss is expressed as

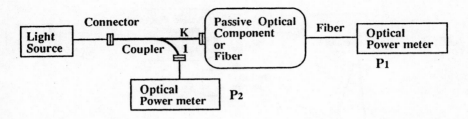

Figure 4.6 Modified insertion method.

$$\alpha = 10 \log_{10} \left[\frac{KP_2}{P_1} \right] \quad \text{(dB)} \qquad (4.8)$$

This method does not require an optical cord, and P_1 and P_2 can be measured simultaneously when using two power meters. This method is especially suitable for measuring loss for a long period, as in the reliability test of components, because the light source power generally fluctuates for a long time and the power monitoring of a light source is always possible with this method.

Several light sources are used for both the cutback and insertion methods: halogen lamp, LED, and laser diode. The attenuation is usually measured in a small-spectrum range and it is expressed as α at wavelength λ. Therefore, a halogen lamp with a monochromator or a halogen lamp with an optical filter is used for a light source. For a stable measurement, light from a halogen lamp is chopped by a chopper and detected by a detector with a lock-in amplifier. Wider dynamic range is realized when using an LED or a laser diode, especially the latter. A stable measurement is realized by using an LED or a laser diode with temperature control.

In the case of loss measurement in multimode fibers, special care must be taken with mode excitation (mode power distribution). Steady-state excitation is commonly used to measure a multimode fiber or a component used for multimode fiber transmission systems because of the high reproducibility of measured values. Another reason for using steady-state excitation may be that the attenuation measured by steady-state excitation is generally closer to the actual loss in the fiber transmission systems than the attenuation measured by other excitations. Strictly speaking, steady-state mode power distribution is different for different fibers, and, therefore, it cannot be realized. Pseudo-steady-state mode power distribution realized by application of a dummy fiber or a mode scrambler is practically used. As an example, and is shown in dummy fiber or a mode scrambler to the modified insertion for a long fiber, Figure 4.7. Since steady-state mode power distribution is realized with a diameter a fiber with a relatively long length, 2 to 3 km, wound on a realized by a bent of about 10 cm is used as a dummy fiber. A mode scram ultimode fibers, such short fiber or by spliced short fibers with different type

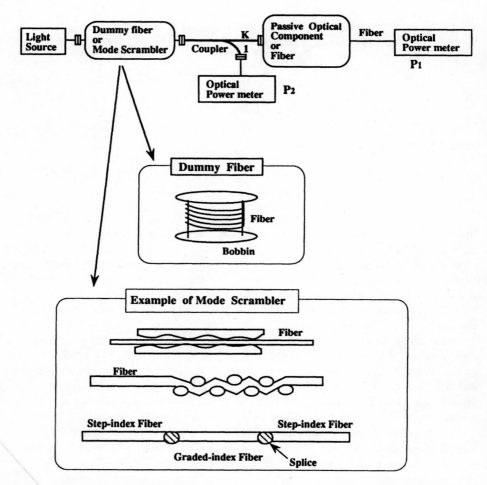

Figure 7 Loss measurement for a multimode fiber or component (application to modified insertion method).

as SI and GI fibers. Either a dummy fiber or a mode scrambler is inserted between a light source and the measured sample (a fiber or a component).

4.2.2 Backscattering Method

The backscattering method is very useful, since it makes it possible to determine a fiber and a component's losses locally and nondestructively. It also useful for fault location. This information (loss, fault location) can be obtained from one end of the fiber; that is, we obtain this information remotely. Online monitoring is possible with the backscattering method with wavelengths other than the transmission signal wavelength in fiber transmission systems. The drawback of this method is that it has less accuracy than the cutback method.

The backscattering method makes use of the scattered light in a fiber to the fiber input end. The origins of scattering are Rayleigh scattering and Fresnel reflection. Rayleigh scattering is one of the major fiber losses, as discussed in Chapter 2, and it is radiated nearly isotropically. Some portions of Rayleigh scattered light propagate backward. The backscattering method is implied by Kapron et al. [19] and was realized in 1976 [20,21]. Fresnel reflection originates the index difference of an interface, such as the air gap connector and fiber break. The reflection coefficient R is given by (3.51) using the plane wave approximation. In the case of an interface between a fiber ($n_A \approx 1.46$) and air ($n_B = 1$), only about 4% of the incident power is reflected. Moreover, the R value of (3.51) is valid for the perpendicular endface reflection. Therefore, the reflected power from a fiber breakpoint is less than 4%, and it may depend on the fiber break shape. Fault location by using this principle [22] will be discussed in Section 4.7.

The backscattered power by Rayleigh scattering is given as

$$P_B = \frac{S \, \alpha_s \, v_g \, W}{2} \, e^{-2\alpha_t L} \, P_{\text{in}} \tag{4.9}$$

where S, α_s, α_t, v_g, W, L, and P_{in} are the backscattering factor, Rayleigh scattering loss, total loss, group velocity, optical pulsewidth in time, fiber length, and input power, respectively. The total loss α_t is a summation of Rayleigh scattering loss and absorption loss. The derivation of (4.9) is shown below by using Figure 4.8. The scattered power $dP_B(z)$ from the infinitesimal fiber length dz (located between z and $z + dz$) is

$$dP_B(z) = S \, dP_S(z) \tag{4.10}$$

where S is defined as the ratio of the Rayleigh backward scattering power to the Rayleigh scattering power. The Rayleigh scattering power from dz is

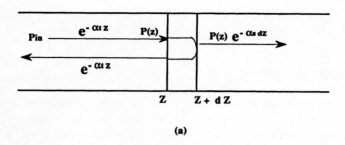

(a)

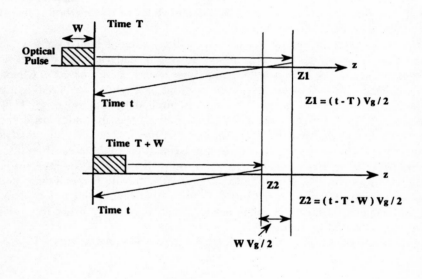

(b)

Figure 4.8 Derivation of backscattered power.

$$dP_S(z) = P(z) - P(z)e^{-\alpha_s} dz \approx P(z)\alpha_s dz \tag{4.11}$$

and the following equation is obtained by combining (4.10) and (4.11):

$$dP_B(z) = SP(z)\alpha_s dz \tag{4.12}$$

When this backscattered power at the fiber input is observed, the power attenuates with $\exp(-\alpha_t z)$; that is,

$$dP_B(z) = SP(z)\alpha_s e^{-\alpha_t z} dz = S\alpha_s e^{-2\alpha_t z} P_{in} dz \tag{4.13}$$

where the relation of $P(z) = P_{in} \exp(-\alpha_t z)$ is used. The return time t from location z to the fiber input endface is expressed as

$$t = \frac{2z}{v_g} \tag{4.14}$$

The backscattered power at time t is then obtained as follows:

$$P_B(t) = \frac{S \alpha_s v_g P_{in}}{2} \int_{t_r}^{t_r+W} e^{-\alpha_t v_g t} dt \approx \frac{S \alpha_s v_g W P_{in}}{2} e^{-\alpha_t v_g t_r} \tag{4.15}$$

This equation is rewritten by using the round-trip time $t_r = 2L/v_g$, and we obtain (4.9). As indicated in Figure 4.8(b), the backscattered light between z and $z + \Delta z$ cannot be distinguished and the overlapped light is integrated in (4.15). The spatial resolution Δz is given as

$$\Delta z = z_1 - z_2 = W v_g/2 \tag{4.16}$$

An example of the spatial resolution Δz is 100 m for $W = 1 \mu s$ and $v_g = 2 \times 10^8$ m.

The backscattering factor S is given by [21,23–25]:

$$S = \frac{3}{8} \frac{n_0^2 - n_2^2}{n_0^2} \quad \text{(SI multimode fiber)} \tag{4.17}$$

$$S = \frac{1}{4} \frac{n_0^2 - n_2^2}{n_0^2} \quad \text{(GI multimode fiber)} \tag{4.18}$$

$$S = \frac{3}{2} \frac{a^2}{\omega^2 V^2} \frac{n_0^2 - n_2^2}{n_0^2} \quad \text{(single-mode fiber)} \tag{4.19}$$

and these are roughly expressed as

$$S = C_r \frac{n_0^2 - n_2^2}{n_0^2} \approx 2C_r \Delta \tag{4.20}$$

where C_r is a constant ranging from 0.2 to 0.4 depending on the fiber type (SI multimode, GI multimode, and single-mode fibers). By combining (4.9) and (4.20), the following equation is obtained.

$$\frac{P_B}{P_{in}} = \alpha_s v_g W C_r \Delta e^{-2\alpha_t L} \tag{4.21}$$

Since Δ for a single-mode fiber is about one-fifth that for a multimode fiber, the backscattering power of a single-mode fiber is less than that of a multimode fiber by a factor of about one-fifth for the same input power. The example of $\alpha_s v_g W C_r \Delta$ is 6.8×10^{-6} (about -52 dB) for $\alpha_s = 0.3$ dB/km, $v_g = 2 \times 10^8$ m/s, $W = 1$ μs, and $C_r \Delta = 5 \times 10^{-4}$ (single-mode fiber). This value is about 38 dB lower when compared with the Fresnel reflection (about -14 dB).

A setup using the backscattering method is called an OTDR, named after the existing electrical time-domain reflectometer. The block diagram of an OTDR is shown in Figure 4.9, where a fiber coupler is used for a directional coupler. Instead of a fiber coupler, a half mirror, a polarization beam splitter, or an acousto-optical (AO) light deflector [26] is also used. Among these, the AO light deflector has superior characteristics, such as low insertion loss, good gating performance, and

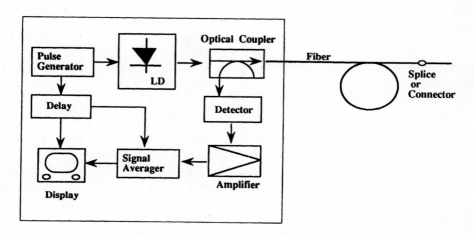

Figure 4.9 OTDR (block diagram of equipment).

low polarization dependence, which are suitable for a single-mode fiber OTDR. The received signals are averaged by adding the N-time measurements in the signal averager block to obtain the high SNR signal. Since the noise is random, the noise power increases by a factor of N, while signal power increases by N^2. The SNR increases in proportion to N for N-time averaging. For a precise measurement, a large number of averaging is inevitable; however, it requires a long measurement time. In an effort to widen the dynamic range of an OTDR, there have been several investigations such as coherent detection [27], photon counting detection [28], the laser diode amplifier [29], and the erbium-doped fiber amplifier [30].

Using (4.21), the ratio of $P_B(z)$ for $z = L_1$ and $z = L_2$ $(L_2 > L_1)$ is expressed as

$$\log_e \left[\frac{P_B(L_1)}{P_B(L_2)} \right] = 2\alpha_t (L_2 - L_1) \tag{4.22}$$

and this expresses the linear relationship between the backscattering power expressed in a logarithm and the fiber length. Fiber loss α_t is measured as a function of fiber length, as shown in Figure 4.10. When a passive optical component, such as a connector, with a loss of α_c, exists between $z = L_1$ and $z = L_2$, the backscattering power ratio is

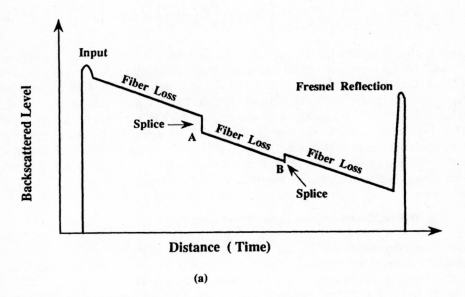

(a)

Figure 4.10 Trace by OTDR: (a) schematic trace; (b) measured trace.

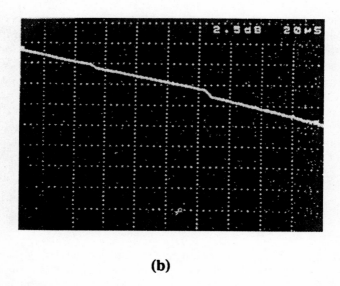

(b)

Figure 4.10 (Continued)

$$\log_e\left[\frac{P_B(L_1)}{P_B(L_2)}\right] \approx \log_e\left[\frac{\alpha_{s1}\,C_{r1}\,\Delta_1}{\alpha_{s2}\,C_{r2}\,\Delta_2}\right] + 2\alpha_c \tag{4.23}$$

where the length difference $(L_1 - L_2)$ is assumed to be small (fiber loss of this portion is small). When input and output fibers of a passive optical component are identical (4.23) is $2\alpha_c$. In this case, the power gap in the trace of Figure 4.10 directly corresponds to the component loss. For the special case of $\alpha_{s1}\,C_{r1}\,\Delta_1/\alpha_{s2}\,C_{r2}\,\Delta_2 < 1$ and $\alpha_c \ll 1$, (4.23) sometimes becomes minus, which is shown as point B in Figure 4.10(a). Usually, the fiber-dependent factor is nearly equal for input and output fibers, and component loss can be roughly evaluated by the OTDR from one fiber endface. The measured trace for spliced fibers is shown in Figure 4.10(b). In this case, splices are clearly seen as gaps like point A in Figure 4.10(a).

4.2.3 Wavelength-Dependent Loss Measurement

Wavelength-dependent loss (or spectral loss) is measured by the cutback or insertion method at several wavelengths. The light source is usually a halogen lamp with a monochromator. Alternatively, a halogen lamp with several optical filters, LEDs, or laser diodes with different emitting wavelengths are also used. By using a monochromator, the continuous measurement of spectral loss is possible. When using

LEDs or laser diodes, only a discreet measurement of spectral loss is possible because of the limited number of emitting wavelengths from LEDs and laser diodes available at present.

4.3 BASEBAND FREQUENCY AND CHROMATIC DISPERSION MEASUREMENT

The baseband frequency response of a fiber is very important for fiber transmission systems, especially high-speed transmission systems. Although the baseband frequency response is important both for single-mode and multimode fiber systems, it is not as important for a passive optical component used in single-mode fiber systems. Since almost all passive optical components, except for fiber-type components that use a long fiber, do not significantly change the baseband frequency response, and since single-mode fibers themselves have a very wide bandwidth, it is not always necessary to measure the baseband frequency response of passive optical components. Practically, chromatic dispersion is measured to evaluate the frequency response instead of baseband frequency response in the case of single-mode fibers. In contrast to components used in single-mode fiber systems, passive optical components used in multimode fiber systems may change the mode power distribution, and this may affect the baseband frequency response. Therefore, the baseband frequency response of the passive optical components with multimode fibers is frequently required to be measured.

4.3.1 Baseband Frequency Measurement in Multimode Fiber Systems

Time-Domain Method

Fibers can be treated as a linear system, as discussed in Chapter 2 [31], and the impulse response of fibers including passive optical components is expressed as

$$P_{out}(t) = \int_{-\infty}^{\infty} h(t - \tau)P_{in}(\tau) \, d\tau \qquad (4.24)$$

where $h(t)$, $P_{in}(t)$, and $P_{out}(t)$ are the impulse response, input pulse power, and output pulse power, respectively. One example of a measurement setup is shown in Figure 4.11, where a directional coupler and a laser diode are used as a beam splitter and a light source, respectively. Light pulse broadening is evaluated from the measured rms pulsewidth W_{rms} for input and output pulse. The rms pulsewidth W_{rms} is defined as

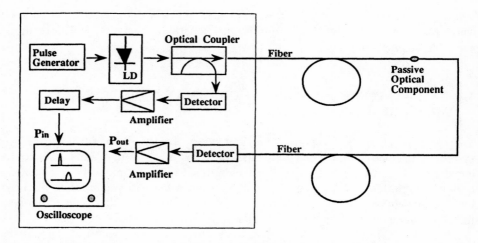

Figure 4.11 Baseband response measurement (time-domain method).

$$W_{\text{rms}}^2 = \int_{-\infty}^{\infty} t^2 P(t)\, \mathrm{d}t - \left[\int_{-\infty}^{\infty} t P(t)\, \mathrm{d}t\right]^2 \qquad (4.25)$$

and the following equation holds using (4.24).

$$
\begin{aligned}
W_{\text{rms out}}^2 &= \int_{-\infty}^{\infty} t^2 P_{\text{out}}(t)\, \mathrm{d}t - \left[\int_{-\infty}^{\infty} t P_{\text{out}}(t)\, \mathrm{d}t\right]^2 \\
&= \int_{-\infty}^{\infty}\int_{-\infty}^{\infty} t^2 h(t-\tau) P_{\text{in}}(\tau)\, \mathrm{d}\tau\, \mathrm{d}t \\
&\quad - \left[\int_{-\infty}^{\infty}\int_{-\infty}^{\infty} t\, h(t-\tau) P_{\text{in}}(\tau)\, \mathrm{d}\tau\, \mathrm{d}t\right]^2 \\
&= \int_{-\infty}^{\infty} t^2 h(t)\, \mathrm{d}t - \left[\int_{-\infty}^{\infty} t\, h(t)\, \mathrm{d}t\right]^2 \\
&\quad + \int_{-\infty}^{\infty} t^2 P_{\text{in}}(t)\, \mathrm{d}t - \left[\int_{-\infty}^{\infty} t\, P_{\text{in}}(t)\, \mathrm{d}t\right]^2 \\
&= W_{\text{rms }h}^2 + W_{\text{rms in}}^2
\end{aligned}
\qquad (4.26)
$$

The impulse response of fibers including passive optical components is measured as $W^2_{rmsh} = W^2_{rms\,out} - W^2_{rms\,in}$. From the standpoint of fiber transmission system design, the baseband frequency response is more convenient than the rms pulse broadening expression. The baseband response is obtained from (4.24) with the Fourier transform; that is,

$$H(\omega) = \frac{P_{out}(\omega)}{P_{in}(\omega)} \qquad (4.27)$$

The baseband response is obtained practically from the measured pulses (input and output pulses) by the FFT method.

Frequency-Domain Method

The frequency-domain method uses (4.27), and a measurement setup is shown in Figure 4.12 as an example. A laser diode is directly frequency-modulated by the sweep generator, and the input and output signals through detectors are measured by a spectrum analyzer. An alternative setup uses one detector; that is, both shaded detector and amplifier in Figure 4.12 are eliminated and the input signal is also detected by one detector, which is indicated by the dashed line. The detected input and output signals are stored in memory and then processed. In this setup, a movable mirror or an optical switch instead of a directional coupler may be suitable for some cases.

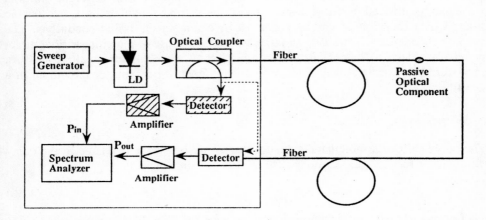

Figure 4.12 Baseband response measurement (frequency-domain method).

4.3.2 Chromatic Dispersion Measurement in Single-Mode Fiber Systems

Chromatic dispersion σ_c is defined in Chapter 2 and again shown here.

$$\sigma_c = \frac{d}{d\lambda}\left(\frac{1}{v_g}\right) = \frac{d\tau}{d\lambda} \tag{4.28}$$

Time-Domain Method

This method is straightforward and is indicated by the definition of σ_c in (4.28). The pulse delay difference Δt for fiber length L is measured with several wavelengths, and the chromatic dispersion σ_c is approximately obtained as

$$\sigma_c \approx \frac{\Delta t/L}{\Delta\lambda} = \frac{\Delta\tau}{\Delta\lambda} = \frac{\tau(\lambda_i) - \tau(\lambda_j)}{\lambda_i - \lambda_j} \tag{4.29}$$

Although the measuring setup for this method is similar to the setup shown in Figure 4.11, there are some modifications. Since the input signal $P_{in}(t)$ is unnecessary in the chromatic dispersion measurement method, the directional coupler, detector, and amplifier used for detection of $P_{in}(t)$ are eliminated from Figure 4.11. The input of an electric pulse from a pulse generated with a delay to an oscilloscope is added in Figure 4.11 for the purpose of a trigger signal. Several candidates for a light source with several wavelengths have been proposed and used, and they are (1) multiple laser diodes, (2) a single tunable laser, including a tunable laser diode, and (3) a fiber Raman laser with a monochromator.

An ordinary laser diode emits a light with one narrow emitting wavelength, and the first candidate uses several ordinary laser diodes with different emitting wavelengths. Instead of using several lasers, the second candidate uses one tunable laser, which can emit several wavelengths by changing the lasing conditions. The third candidate uses a fiber Raman laser, which makes use of a nonlinear fiber effect positively. Since a fiber Raman laser emits many wavelengths simultaneously, a monochromator or an optical filter must be set either at the input or the output of a test fiber.

Frequency-Domain Method

The phase shift of a modulated signal at the output of a fiber is related to the delay difference τ for one guided mode. The phase shift in a fiber is

$$\phi = 2\pi f_m t = 2\pi f_m L/v_g = 2\pi f_m L\tau \tag{4.30}$$

where f_m and L are the modulation frequency and the fiber length, respectively. By using the relation of (4.30), the chromatic dispersion σ_c is measured by the phase shift difference $\Delta\phi$ for different wavelengths [32].

$$\sigma_c = \frac{\Delta\tau}{\Delta\lambda} = \frac{\Delta\phi}{2\pi f_m L \Delta\lambda} \qquad (4.31)$$

The phase shift is determined by observing a Lissajous figure on an oscilloscope display (used in [32]), or by an ordinary vector voltmeter.

The chromatic dispersion σ_c is roughly measured by the baseband frequency response through (4.32) for a single-mode fiber, which was discussed in (2.46).

$$\sigma_c \approx \frac{0.55}{f_{3dB} L \delta\lambda} \qquad (4.32)$$

This equation cannot be applicable to a multimode fiber because of the modal dispersion. For measuring the chromatic dispersion of a multimode fiber as well as a single-mode fiber, a variable light source spectrum method using an LED and an optical filter has been proposed [33]. First, the baseband response is measured by modulating an LED in an ordinary method. Next, the baseband response is again measured using an optical filter placed at the output of a fiber. The optical filter is used for narrowing the light source spectrum width. By taking the ratio of two measuring results and assuming the Gaussian spectrum of an LED, only the chromatic dispersion is measured.

Interferometric Method

The chromatic dispersion σ_c for a very short fiber (e.g., 1 m) is measured by interferometry [34]. The measuring setup is shown in Figure 4.13, and a fiber Raman laser or laser diodes are used as a light source. The lights of two optical paths interfere on the detector (vidicon) plane and visibility V is measured. Visibility V is defined as [35]

$$V = \frac{I_{max} - I_{min}}{I_{max} + I_{min}} \qquad (4.33)$$

where I_{max} and I_{min} are the maximum and minimum intensities of the interference fringes on the detector. The measuring procedure takes two steps. First, the length of the air path is adjusted in order to obtain the maximum visibility for a light source with wavelength λ_1. Next, after changing wavelength λ_2, the length of the air path is readjusted in order to obtain maximum visibility. When the path length change by readjustment is $2d$, the time difference for two measuring steps is $[\tau(\lambda_1) - \tau(\lambda_2)]L = 2d/c$, where L and c are the fiber length and the speed of light, respectively. Therefore, the chromatic dispersion σ_c is given by

$$\sigma_c = \frac{\Delta\tau}{\Delta\lambda} = \frac{2d}{c L(\lambda_1 - \lambda_2)} \qquad (4.34)$$

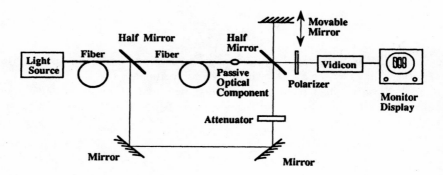

Figure 4.13 Interferometer for measuring chromatic dispersion.

4.4 REFLECTION PROPERTIES MEASUREMENT

Evaluation of the reflection due to passive optical components is very important, because the reflection may have a serious effect on the transmission properties in some fiber transmission systems, as discussed in Chapter 3.

Reflection is simply measured by using an optical directional coupler as shown in Figure 4.14 [36]. The stabilized light source and a good optical directional coupler with low crosstalk (high directivity) are required in this method. The reflected powers for no reflection, perfect reflection (100% reflection), and a test component are measured as P_0, P_1, and P_2, respectively. Reflection R of a test component is given as

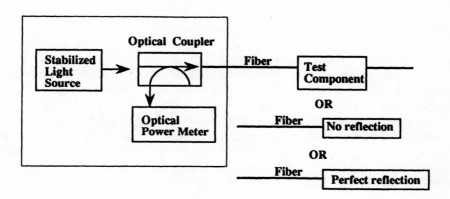

Figure 4.14 Reflection measuring method using coupler.

$$R = \frac{P_2 - P_0}{P_1 - P_0} \qquad (4.35)$$

No reflection and perfect reflection are approximately realized by using a matching material and a mirror, respectively. The detectable reflection with this method depends on the performance of an optical directional coupler, and it is about -40 to -50 dB. Reflection under -50 dB is difficult.

Reflection can be measured by an OTDR [37]. The OTDR trace contains both reflection by Rayleigh scattering and reflection by passive optical components, as shown in Figure 4.15. These can be evaluated separately. As discussed in the previous section of this chapter (4.21), reflection by Rayleigh scattering is expressed as

$$R_{\text{Rayleigh}} = \alpha_s v_g W C_r \Delta \qquad (4.36)$$

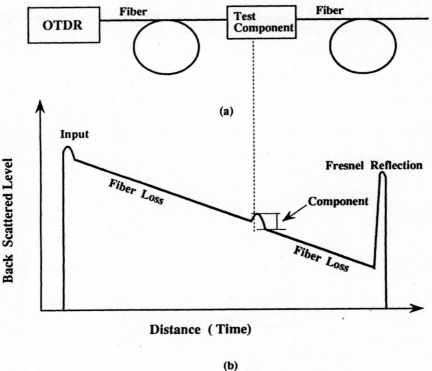

Figure 4.15 Reflection measuring method using OTDR: (a) setup; (b) schematic trace.

and this value can be used as the reference value to evaluate the reflection of a test component. The detectable reflection with this method is about -50 to -60 dB. Since α_s is smaller for longer wavelengths, R_{Rayleigh} becomes smaller. Therefore, the detectable reflection tends to become lower for longer wavelengths.

4.5 FIBER ENDFACE QUALITY MEASUREMENT

The fiber endface influences connection loss and reflection properties. According to [38], a connector with a rough polishing surface made by a 15-μm grain size has about 0.5 dB of additional connection loss when compared to that with a fine polishing surface made by a 1-μm grain size. A good fiber endface is also important for a fusion splice to obtain low splice losses. Several measuring methods are shown in Figure 4.16.

The method shown in Figure 4.16(a) uses a photodiode with four electrodes [39]. The light from a fiber attached on a manipulator projects on this photodiode, and the voltages V_x and V_y are detected. Then the fiber is rotated by 180 deg and again the light is detected. By moving the manipulator (Δx, Δy), the same values of V_x and V_y are obtained. The fiber endface angle θ_f is measured by

$$\theta_f = \tan^{-1}\left(\frac{\sqrt{(\Delta x)^2 + (\Delta y)^2}}{2L}\right) \tag{4.37}$$

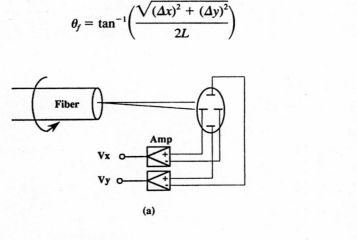

(a)

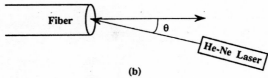

(b)

Figure 4.16 Fiber endface quality measurement.

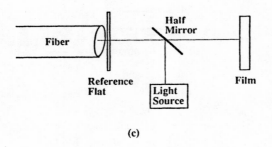

(c)

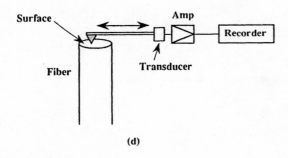

(d)

Figure 4.16 (Continued)

where L is the distance between a fiber and a photodiode. The reflection method using a He-Ne laser is shown in Figure 4.16(b), where the reflected light from a fiber is detected and the angle θ_f is measured [40]. The third method, shown in Figure 4.16(c), uses the interferometric technique. The fiber endface quality can be evaluated from the interferometric patterns [38,41]. From the fringe interval, connector surface angle distribution can be measured [42]. The fourth method, shown in Figure 4.16(d), is the mechanical method, where the surface roughness is directly measured through the contact of a fiber surface and a needle.

4.6 OPTICAL CROSSTALK MEASUREMENT

Optical crosstalk is important when two fibers are closely aligned, such as a fiber ribbon and an array connector. The setup for measuring the crosstalk is shown in Figure 4.17, where the optical power meter with two ports (two detectors) is used. It is possible to use an ordinary one-port power meter. The crosstalk is measured by the detected optical powers P_s and P_x; for example,

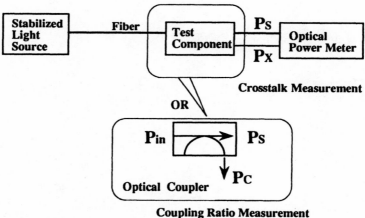

Figure 4.17 Crosstalk and coupling ratio measurement.

$$\text{Crosstalk} = 10 \log \frac{P_x}{P_s} \tag{4.38}$$

The insertion loss and coupling ratio of a directional coupler are measured by using the same setup. They are

$$\text{Insertion loss} = 10 \log \frac{(P_s + P_c)}{P_{\text{in}}} \tag{4.39}$$

and the coupling ratio is P_c/P_s.

4.7 FAULT LOCATION AND DIAGNOSIS MEASUREMENT

Fault location, such as the fiber breakpoint, can be measured by using the Fresnel reflection. The Fresnel reflection is schematically shown in Figure 4.10(a). As already discussed, the reflected power is less than 4% of the incident power, and this value is large when compared to the Rayleigh backscattering power. Ordinary OTDR is useful for detecting the fiber breakpoint in fiber transmission systems. However, it is not sufficient to detect the breakpoint or the local loss of passive optical components, because higher spatial resolution is required. Spatial resolution on the order of submicrons or microns is required for the diagnosis of passive optical components. Several methods used for the diagnosis have been proposed, classified into (1) time-domain measurement, (2) frequency-domain measurement, and (3) low coherence and interferometric measurement.

The OTDR with high spatial resolution has been proposed for the time-domain measurement. One early proposed method uses a high-power dye laser (about 300W and wavelength of 0.6 μm) with a short pulse (0.1 to 0.3 ps) and nonlinear detection [43]. The setup is shown in Figure 4.18. The laser beam is split into reference and probe beams by a half mirror (a beam splitter). The reference beam is reflected by a prism and then focused into a KH$_2$PO$_4$ (KDP) crystal. The backscattered probe beam from a passive optical component is also focused into the KDP crystal. Due to the nonlinearity of the KDP crystal, the light with a double frequency (0.3-μm wavelength in this case) is generated. The generated power of the frequency-doubling light is proportional to the product of two beams and is measured by a detector with an optical filter. By scanning the reference beam (by moving the prism position),

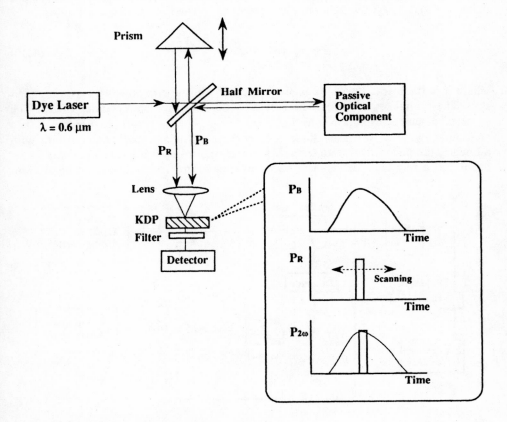

Figure 4.18 OTDR with nonlinear detection. (After [43].)

the backscattered power is measured as a function of the component position. The reported spatial resolution is about 15 μm.

Many methods have been proposed for the frequency-domain measurement [44–50]. Among them, two are explained here using Figures 4.19 and 4.20. In Figure 4.19, an optical source such as a laser diode is intensity-modulated with a swept modulation frequency. The electric signal used for the modulation has a constant amplitude with a periodic linear frequency sweep. The backscattered optical signal from a passive optical component is detected and amplified. The amplified signal is mixed with the source drive signal in order to produce a correlation signal in the frequency domain. The frequency difference between the backscattered signal and the source drive signal corresponds to the delay caused by propagation in the passive optical component. Therefore, the frequency axis of the spectrum analyzer is proportional to the distance in the passive optical component. We assume that the swept frequency ranges from $f_c - f_d$ to $f_c + f_d$ and a repetition frequency is f_s. The delay is $\Delta f/(2 f_d f_s)$ for frequency difference Δf. Then the distance x is [44]

$$x = v_g \cdot (\text{delay})/2 \approx \frac{c \, \Delta f}{4n \, f_s f_d} \tag{4.40}$$

where n is the refractive index of the waveguide in the component.

The spatial resolution is determined by the measurement accuracy of Δf. The reported distance resolution is 3.5 m in the experiment in [44]. In Figure 4.20, a continuous-wave (CW) laser diode is frequency-modulated to produce a light with a linearly ramped optical frequency. The input beam is split into two beams, a probe beam and a reference beam. The backscattered beam from a passive optical com-

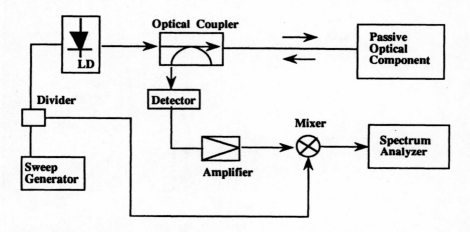

Figure 4.19 Frequency-domain reflectometry (IM). (After [44].)

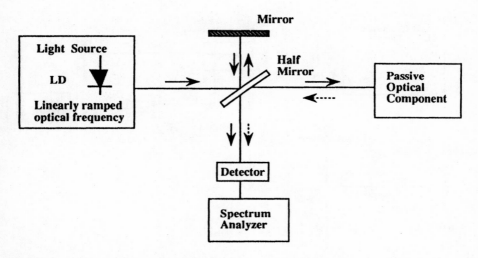

Figure 4.20 Frequency-domain reflectometry (FM).

ponent and the reference beam with a constant delay are mixed at a detector. The different propagation times result in different optical frequencies (wavelengths) because of the light source with a linearly ramped optical frequency. The mixed beams produce a beat frequency that corresponds to the delay caused by propagation in the passive optical component. Therefore, the frequency axis of the spectrum analyzer is proportional to the distance in the passive optical component. The spatial resolution Δx is determined by [50]

$$|\Delta x| = \left| \frac{\lambda^2}{2n \, \delta\lambda} \right| = \left| \frac{c}{2n \, \delta f} \right| \qquad (4.41)$$

where $\delta\lambda$ and δf are the wavelength shift and the optical frequency shift of the light source, respectively. For example, if $\delta f = 100$ GHz, then the spatial resolution $\Delta x = 1$ mm for $n = 1.5$. A spatial resolution of $\Delta x = 50$ μm and a 100-dB dynamic range have been reported [50].

The hybrid methods using both time- and space-domain measurements use an interferometer and a light source with a short coherent length (a low-coherence light source) [51–53]. A superluminescent diode (SLD), a laser diode biased below threshold, and an LED have been used as low-coherence light sources. These methods use a CW light source in contrast to the usual OTDR method, which uses a short light pulse. One proposed method among them is shown in Figure 4.21 [52]. In this case, an SLD with a coherent length of $Lc = 50$ μm is used as a low-coherence light

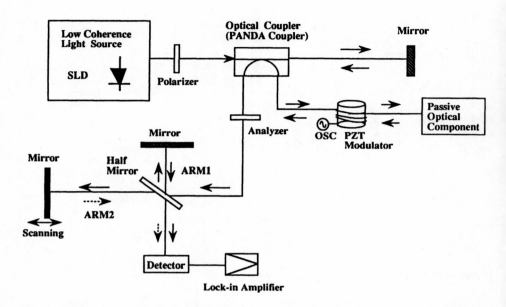

Figure 4.21 Low coherence and interferometric reflectometry. (After [52].)

source. The output light from the SLD is launched into the polarization-maintaining fiber coupler (PANDA coupler), which serves as an interferometer. This construction ensures low polarization noises. The light is divided into two beams by the PANDA coupler. One beam, which is phase-modulated by a piezoelectric transducer (PZT) modulator, enters a passive optical component under test, and the other beam reflected by a mirror is used as a reference light. The combination of the phase-modulation scheme and the lock-in amplifier ensures that there are stable and wide dynamic range measurements. The phase-modulation frequency is 6 kHz in this case. The backscattered and reference lights enter the second interferometer. In this second interferometer, the two lights interfere and then are detected. The detected signal is measured by the lock-in amplifier. By moving the mirror in this second interferometer, the backscattered light from the passive optical component can be measured as a function of the position in the component. The optical path difference ΔL of two scattering points is expressed as

$$\Delta L = 2n \, \Delta s \tag{4.42}$$

where n and Δs are the refractive index of the waveguide in the component under test and the distance between two scattering points in the component. Factor 2 represents the forward and backscattered propagation distance in the component. When the optical path difference satisfies the inequality of $\Delta L > Lc$ (Lc = coherent length), then two backscattered lightwaves independently interfere with the reference lightwave. Therefore, the spatial resolution Δx is

$$\Delta x = Lc/2n \tag{4.43}$$

and

$$\Delta x = \frac{50}{2 \times 1.5} \approx 17 \ \mu m$$

for the parameter in this experiment. The actual resolution is degraded in accordance with the averaging to improve the SNR of the detected signal. The realized spatial resolution has been reported to be about 400 μm. A superluminescent light source using an erbium-doped fiber and a balanced mixer detection has been used to improve the minimum detectable reflectivity, and a minimum detectable reflectivity value of -146 dB has been demonstrated [54].

REFERENCES

[1] Streckert, J., "New Method for Measuring the Spot Size of Single-Mode Fibers," Opt. Lett., Vol. 5, 1980, p. 505.

[2] Alard, F., L. Jeunhomme, and P. Sansonetti, "Fundamental Mode Spot-Size Measurement in Single-Mode Optical Fibers," Electron. Lett., Vol. 17, 1981, p. 958.

[3] Di Vita, P. P., G. Coppa, and U. Rossi, "Characterization Methods for Single-Mode Fibers," 10th European Conf. on Optical Communications (ECOC'84), 1984, p. 48.

[4] Martin, W. E., "Refractive Index Profile Measurements of Diffused Optical Waveguides," Appl. Opt., Vol. 13, 1974, p. 2112.

[5] Cherin, A. H., L. G. Cohen, W. S. Holden, C. A. Burrus, and P. Kaiser, "Transmission Characteristics of Three Corning Multimode Optical Fibers," Appl. Opt., Vol. 13, 1974, p. 2359.

[6] Wonsiewicz, B. C., W. G. French, P. D. Lazay, and J. R. Simpson, "Automatic Analysis of Interferograms: Optical Waveguide Refractive Index Profiles," Appl. Opt., Vol. 15, 1976, p. 1048.

[7] Gloge, D., and E. A. J. Marcatili, "Multimode Theory of Graded-Core Fibers," Bell Syst. Tech. J., Vol. 52, 1973, p. 1563.

[8] Adams, M. J., D. J. Payne, and F. M. E. Sladen, "Leaky Rays on Optical Fibers of Arbitrary (Circularly Symmetric) Index Profiles," Electron. Lett., Vol. 11, 1975, p. 238.

[9] Ikeda, M., M. Tateda, and H. Yoshikiyo, "Refractive Index Profile of a Graded Index Fiber: Measurement by a Reflection Method," Appl. Opt., Vol. 14, 1975, p. 814.

[10] Eickhoff, W., and E. Weidel, "Measuring Method for the Refractive Index Profile of Optical Glass Fibers," Opt. & Quant. Electron., Vol. 7, 1975, p. 109.

[11] Stewart, W. J., "A New Technique for Measuring the Refractive Index Profiles of Graded Optical Fibers," Int. Conf. Integrated Opt. Fiber Commun. (IOOC), 1977, p. 395.

[12] White, K. I., "Practical Application of the Refracted Near-Field Technique for the Measurement of Optical Fiber Refractive Index Profiles," Opt. & Quant. Electron., Vol. 11, 1979, p. 185.

[13] Marhic, M. E., P. S. Ho, and M. Epstein, "Nondestructive Refractive-Index Profile Measurements of Clad Optical Fibers," Appl. Phys. Lett., Vol. 26, 1975, p. 574.

[14] Iga, K., Y. Kokubun, and N. Yamamoto, "Refractive Index Profile Measurement of Focusing Fibers by Using a Transverse Differential Interferometry," IECE of Japan, Technical Report OQE76-80, 1976 (in Japanese).

[15] Kokubun, Y., and K. Iga, "Precise Measurement of the Refractive Index Profile of Optical Fibers by a Nondestructive Interference Method," IECE of Japan, Vol. E60, 1977, p. 702.

[16] Okoshi, T., and K. Hotate, "Computation of the Refractive Index Distribution in an Optical Fiber From Its Scattering Pattern for a Normally Incident Laser Beam," Opt. & Quant. Electron., Vol. 8, 1976, p. 78.

[17] Okoshi, T., and K. Hotate, "Refractive-Index Profile of an Optical Fiber: Its Measurement by the Scattering-Pattern Method," Appl. Opt., Vol. 15, 1976, p. 2756.

[18] Saekeng, C., and P. L. Chu, "Nondestructive Determination of Refractive-Index Profile of an Optical Fiber: Forward Light Scattering Method," Electron. Lett., Vol. 14, 1978, p. 802.

[19] Kapron, F. P., R. D. Maurer, and M. P. Teter, "Theory of Backscattering Effects in Waveguide," Appl. Opt., Vol. 11, 1972, p. 1352.

[20] Barnoski, M. K., and S. M. Jensen, "Fiber Waveguide: A Novel Technique for Investigating Attenuation Characteristics," Appl. Opt., Vol. 15, 1976, p. 2112.

[21] Personick, S. D., "Photon Probe—An Optical-Fiber Time-Domain Reflectometer," Bell Syst. Tech. J., Vol. 56, 1977, p. 355.

[22] Ueno, Y., and M. Shimizu, "Optical Fiber Fault Location Method," Appl. Opt., Vol. 15, 1976, p. 1385.

[23] Neumann, E.-G., "Analysis of the Backscattering Method for Testing Optical Fiber Cable," Vol. 34, 1980, p. 157.

[24] Brinkmeyer, E., "Analysis of the Backscattering Method for Single-Mode Optical Fibers," J. Opt. Soc. Am., Vol. 70, 1980, p. 1010.

[25] Nakazawa, M., "Rayleigh Backscattering Theory for Single-Mode Optical Fibers," J. Opt. Soc. Am., Vol. 73, 1983, p. 1175.

[26] Nakazawa, M., T. Tanifuji, M. Tokuda, and N. Uchida, "Photon Probe Fault Locator for Single-Mode Optical Fiber Using an Acoustooptical Light Deflector," IEEE J. Quantum Electron., Vol. QE-17, 1981, p. 1264.

[27] Healey, P., and D. J. Malyon, "OTDR in Single-Mode Fiber at 1.5 μm Using Heterodyne Detection," Electron. Lett., Vol. 18, 1982, p. 862.

[28] Healey, P., and P. Hensel, "Optical Time Domain Reflectometry by Photon Counting," Electron. Lett., Vol. 16, 1980, p. 631.

[29] Suzuki, K., T. Horiguchi, and S. Seikai, "Optical Time Domain Reflectometer With a Semiconductor Laser Amplifier," Electron. Lett., Vol. 20, 1984, p. 714.

[30] Blank, L. C., and D. M. Spirit, "OTDR Performance Enhancement Through Erbium Fiber Amplification," Electron. Lett., Vol. 25, 1989, p. 1693.

[31] Personick, S. D., "Time Dispersion in Dielectric Waveguides," Bell Syst. Tech. J., Vol. 50, 1971, p. 843.

[32] Daikoku, K., and A. Sugimura, "Direct Measurement of Wavelength Dispersion in Optical Fibers Difference Method," Electron. Lett., Vol. 14, 1978, p. 149.

[33] Tanifuji, T., and M. Ikeda, "Simple Method for Measuring Material Dispersion in Optical Fibers," Electron. Lett., Vol. 14, 1978, p. 367.

[34] Tateda, M., N. Shibata, and S. Seikai, "Interferometric Method for Chromatic Dispersion Measuring in a Single-Mode Optical Fiber," IEEE J. Quantum Electron., Vol. QE-17, 1981, p. 404.

[35] Born, M., and E. Wolf, Principle of Optics, Pergamon Press, 1975.

[36] Suzuki, N., M. Saruwatari, and M. Okuyama, "Low Insertion- and High Return-Loss Optical Connectors With Spherically Convex-Polished End," Electron. Lett., Vol. 22, 1986, p. 110.

[37] Sankawa, I., T. Satake, N. Kashima, and S. Nagasawa, "Fresnel Reflection Reducing Methods for Optical-Fiber Connector With Index Matching Materials," IECE of Japan, J67-B, 1984, p.

1423. (In Japanese.) English translation is in Electronics and Communications in Japan, Part 1, Vol. 69, Scripta Technica, Inc., 1986, p. 94.

[38] Tuchiya, H., H. Nakagome, N. Shimizu, and S. Ohara, "Double Eccentric Connectors for Optical Fibers," Appl. Opt., Vol. 16, 1977, p. 1323.

[39] Fujii, Y., and N. Suzuki, "Precise Angular Misalignment Measurement in Optical Fiber Connector Plugs," Rev. of Elect. Commun. Lab. (Japan), Vol. 27, 1978, p. 1881 (in Japanese).

[40] Millar, C. A., "A Measurement Technique for Optical Fiber Break Angles," Opt. and Quant. Electron., Vol. 13, 1981, p. 125.

[41] Gordon, K. S., E. G. Rawson, and A. B. Nafarrate, "Fiber-Break Testing by Interferometry: A Comparison of Two Breaking Methods," Appl. Opt., Vol. 16, 1977, p. 818.

[42] Sankawa, I., N. Kashima, and T. Satake, "Testing Method for Optical-Fiber-Connector Surfaces," IECE of Japan, Vol. J67-B, 1984, p. 226 (in Japanese).

[43] Fontaine, J. J., J.-C. Diels, C.-Y. Wang, and H. Sallaba, "Subpicosecond-Time-Domain Reflectometry," Opt. Lett., Vol. 6, 1981, p. 405.

[44] MacDonald, R. I., "Frequency Domain Optical Reflectometer," Appl. Opt., Vol. 20, 1981, p. 1840.

[45] Eickhoff, W., and R. Ulrich, "Optical Frequency Domain Reflectometry in Single-Mode Fiber," Appl. Phys. Lett., Vol. 39, 1981, p. 693.

[46] Kingsley, S. A., and D. E. N. Davies, "OFDR Diagnostics for Fiber and Integrated-Optic Systems," Electron. Lett., Vol. 21, 1985, p. 434.

[47] Ghafoori-Shiraz, H., and T. Okoshi, "Optical-Fiber Diagnosis Using Optical-Frequency-Domain Reflectometry," Opt. Lett., Vol. 10, 1985, p. 160.

[48] Uttam, D., and B. Culshaw, "Precision Time Domain Reflectometry in Optical Fiber Systems Using a Frequency Modulated Continuous Wave Ranging Technique," IEEE J. Lightwave Technol., Vol. LT-3, 1985, p. 971.

[49] Nakayama, J., K. Iizuka, and J. Nielsen, "Optical Fiber Fault Locator by the Step Frequency Method," Appl. Opt., Vol. 26, 1987, p. 440.

[50] Barfuss, H., and E. Brinkmeyer, "Modified Optical Frequency Domain Reflectometry With High Spatial Resolution for Components of Integrated Optic Systems," IEEE J. Lightwave Technol., Vol. 7, 1989, p. 3.

[51] Youngquist, R. C., S. Carr, and D. E. N. Davies, "Optical Coherence-Domain Reflectometry: A New Optical Evaluation Technique," Opt. Lett., Vol. 12, 1987, p. 158.

[52] Takada, K., I. Yokohama, K. Chida, and J. Noda, "New Measurement System for Fault Location in Optical Waveguide Devices Based on an Interferometric Technique," Appl. Opt., Vol. 26, 1987, p. 1603.

[53] Danielson, B. L., and C. D. Whittenberg, "Guided-Wave Reflectometry With Micrometer Resolution," Appl. Opt., Vol. 26, 1987, p. 2836.

[54] Takada, K., M. Shimizu, M. Yamada, M. Horiguchi, A. Himeno, and K. Yukimatu, "Ultrahigh-Sensitivity Low Coherence OTDR Using Er^{3+}-Doped High-Power Superfluorescent Fiber Source," Electron. Lett., Vol. 28, 1992, p. 29.

PART II
SIMPLE CONNECTION

Chapter 5
Fusion Splicing by Discharge

There are several proposed fusion-splicing methods, all of which are classified by heating sources: an electric discharge, a gas-laser, and a flame. In this chapter, the method of fusion-splicing by an electric discharge is explained. This splicing method is widely used for splicing silica fibers. Several important techniques for the electric-discharge fusion-splicing method, such as the prefusion method, the high-frequency discharge with high-voltage trigger method (the HHT method), and the uniform-heating method for mass-fusion splice, have been developed and they are explained here. Also explained are fiber-alignment methods for fusion splicing and methods of protecting spliced fibers.

5.1 OUTLINE AND HISTORY OF FUSION SPLICING

The fundamental requirements for fiber connection methods, including fusion-splicing, are shown below:

1. Little degradation of fiber transmission properties (low connection loss and small reflection);
2. High reliability of connection;
3. Good handling or workability even in the field;
4. Low cost of connection.

There has been a great deal of research on fusion splices, and through intensive research, fusion-splicing methods now satisfy all these requirements. Fusion-splicing methods use a heat source to melt fibers. Unlike other methods using matching materials or adhesives, no materials exist other than the fiber itself in the spliced part. Therefore, the fusion splice has inherently low reflection and high reliability. The early stage of fusion splice research is listed in Table 5.1. In 1971, D. L. Bisbee used an electric heater with a nichrome wire as a heat source to splice composite-glass fibers [1]. This may have been the first experiment on the fusion-splicing method.

Table 5.1
Early Fusion Splice Research

Fiber (Melting Point)	Fusion Method	Researcher	Year Published	Reference
Composite-glass fiber (about 600°C)	Heating by nichrome wire	Bisbee	1971	[1]
Silica fiber (about 2,000°C)	Heating by CO$_2$ laser	Fujita et al.	1975	[2]
	Heating by discharge	Hatakeyama et al.	1975	[3]
		Kohanzadeh	1976	[4]
	Heating by microburner	Joctur and Tarday	1976	[5]

Composite-glass fibers have a low melting temperature, about 600°C. The fibers used for today's optical transmission are silica-based, and their melting temperature is about 2,000°C. The temperature generated by the nichrome heater is not sufficient to melt silica-based fibers. For silica fibers, a heating method using a CO$_2$ laser was proposed by Fujita et al. in 1975 [2]. As an alternative method, an electric discharge as a heat source was proposed by Hatakeyama and Tsuchiya [3] and Kohanzadeh [4]. The other alternative method, using the flame of a microburner, was proposed by Jocteur and Tarday [5]. The fusion-splicing method using a discharge has been widely used as a fusion-splicing method, mainly because a compact fusion-splicing machine with easy operation was realized, which is essential for field use.

The process of fusion splicing using a discharge is shown in Figure 5.1. First, fiber coatings are removed and fibers are cut. Then two fibers are set with a gap on a fusion-splicing machine and a start button is pushed. The process from the removal of coatings to pushing a start button is undertaken by an operator (i.e., manually). After the start, fibers are moved to decrease the gap. During the movement of fibers, an electric discharge is generated and maintained for a predetermined period. This part of the process takes place automatically in today's ordinary fusion-splicing machines. Finally, the fusion-spliced part is protected for easy handling. Recently, a fully automatic machine has been developed, which performs all actions automatically from removal to protection. Many investigations have been undertaken to realize these practical fusion-splicing machines using an electric discharge. Some of the most important research projects are listed in Table 5.2. These technologies are explained in detail in the following sections of this chapter. Here, the meaning and the effects of these technologies are explained. The prefusion method is applied to fibers with a nonperfect endface quality, and it realizes low-loss splices even for the nonperfect fiber endfaces [6]. Fibers cut in the field or cut by unskilled operators have endfaces of nonperfect quality; therefore, the prefusion method is essential to

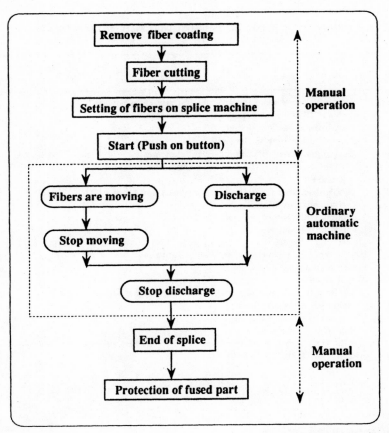

Figure 5.1 Process of fusion splicing.

obtain low-loss splices. The effect of surface tension during a discharge has been investigated, and it works on two mated fibers for the self-alignment of the fiber outer diameter [7]. This effect results in low-loss splices for accurately fabricated fibers, especially single-mode fibers. However, the effect is harmful for core-alignment methods, which will be explained in Section 5.5. The HHT method (high-frequency discharge with high-voltage trigger method) makes the discharge stable for a low-loss splice and realizes a compact fusion-splicing machine with its low power consumption [8]. The effects of the prefusion and HHT methods are shown in Figure 5.2 for splice loss and power consumption. These two technologies are commonly

Table 5.2
Fundamental Research on Practical Fusion-Splicing Machine Using Discharge

Item	Effects	Researcher	Year Published	Reference
Prefusion method	Low-loss splice even for mediocre fiber endface quality; realizes a practical fusion-splicing machine in the field	Hirai and Uchida	1977	[6]
Surface tension effect	Self-alignment of fiber outer diameter due to surface tension	Hatakeyama and Tsuchiya	1978	[7]
HHT method*	Stable discharge for mass-fusion splice by high-frequency discharge; reduces power consumption by high-voltage trigger	Kashima and Nihei	1981	[8]
Offset heating method for mass-fusion splice	Uniform heating for mass-fusion splice using two pairs of electrodes	Tachikura	1981	[9]
	Uniform heating for mass-fusion splice using a pair of electrodes (using high-frequency discharge)	Tachikura and Kashima	1984	[10]

*High-frequency discharge with high-voltage trigger.

used for a single-fusion splice and mass-fusion-splicing methods and for both single-mode and multimode fiber splices. The offset heating method has been developed for mass-fusion splices [9,10]. The mass-fusion splice, shown in Figure 5.3, makes multiple fibers (eight fibers in the figure) melt and splices them simultaneously with one discharge. For this purpose, a uniform heating of multiple fibers must be realized. From the investigation of the heat distribution of an electric discharge, the offset heating method was invented. First, a method using two pairs of electrodes (four electrodes) with a low-frequency discharge (50 Hz) was proposed [9]. However, this method is not suitable for a field-use machine, because there must be accurate alignment of four electrodes, which takes time and skill. Later, a method using a pair of electrodes (two electrodes) with a high-frequency discharge (using the HHT method) was proposed [10]. This method realizes a practical mass-fusion-splicing machine for field use. Up to 12 fibers can be mass-spliced with today's mass-fusion-splicing machines.

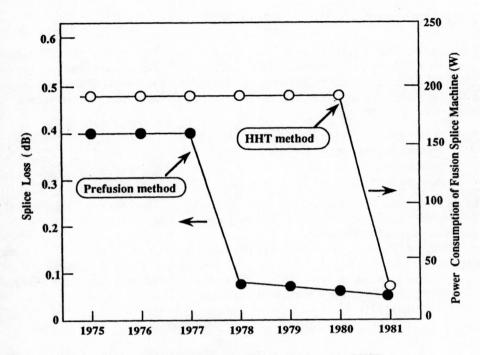

Figure 5.2 History of fusion-splicing machine development (in the case of NTT).

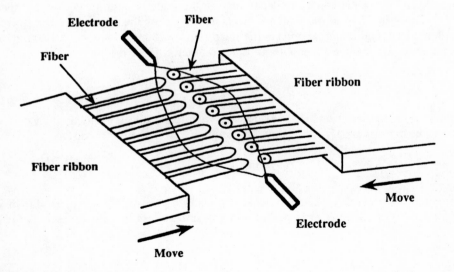

Figure 5.3 Mass-fusion splice.

5.2 FIBER-CUTTING AND PREFUSION METHOD

Fiber endface quality affects splice loss, and polished fiber endfaces are difficult to use with a fusion-splicing method. Therefore, several fiber-cutting methods that do not use the polishing machine have been proposed [11–15]. In four of these methods, a fiber is scored and then bent. These fiber-scoring methods are classified according to the kind of score:

1. A score is made by a blade [11].
2. A score is made by an electric discharge [12].
3. A score is made by a hot wire [14].
4. A score is made by a CO_2 laser [15].

In the fifth method, a fiber is scored and pulled straight without any bending [13]. Among these methods, the score-and-bend method with a blade or other suitable hardware is widely used today.

Glass fibers break in such a way that the fractured endfaces consist of three regions: the mirror, mist, and hackle regions. The mirror region is a smooth surface and is suitable for fusion splices, since flat and perpendicular endfaces are required for splicing. It is known that the mirror region is obtained when the following equation is satisfied [11].

$$\sigma\sqrt{r} = K \tag{5.1}$$

where K and σ are the constants depending on the material and local stress, respectively. The r in (5.1) is the distance from the origin of the fracture to the mirror-mist boundary, and it corresponds to the mirror width. To generate the mirror region, the required stress for a fiber with an outer diameter of $2d$ is

$$\sigma_0 = K/\sqrt{2d} \tag{5.2}$$

When $\sigma > \sigma_0$, the mist or hackle region is generated. A lip is formed when $\sigma < 0$. Therefore, σ must satisfy the following inequality over the entire surface.

$$0 < \sigma < \sigma_0 \tag{5.3}$$

The model of a crack in a sheet, which is made by an elastic body, is shown in Figure 5.4. When this sheet is pulled with stress σ, stress σ_m is expressed as [16]

$$\sigma_m = 2\sigma \sqrt{\frac{c}{\rho}} \quad (\rho \ll c) \tag{5.4}$$

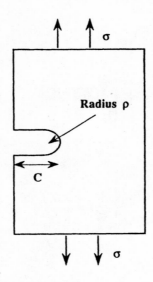

Figure 5.4 Model of an elastic sheet with a crack.

Here, the crack is assumed to have a semicircular tip with a radius ρ and a depth c. The original fracture should be small when obtaining the required stress described by the inequality (5.3). From this equation, ρ must be small; that is, the original crack must be sharp. Using this principle, several cutting tools have been developed and are available. An automatic cutting tool has also been developed [17].

Apart from the common splice loss factors, such as a lateral displacement and a tilt, the unique loss factor for the fusion-splicing method is a bubble confined at the fiber-to-fiber boundary. When bubbles are confined, they result in very large splice loss of a few decibels or more. In a splice procedure used before the proposal of the prefusion method, fibers to be spliced are pressed softly and then fused together by an electric discharge. When the fiber endfaces are not smooth, lateral displacement and bend occur between the spliced fibers in the previous method. The probability of bubble generation is also high, which may be the result of air confinement of rough surfaces. To overcome these difficulties, the prefusion method was proposed [6]. The procedure is shown in Figure 5.5. After setting the fibers with an end separation of several microns, fiber ends are prefused by a discharge (Fig. 5.5 (b)). Smooth surfaces are obtained with this prefusion. Then they are moved and pressed together under the discharge. After the endfaces touch, fibers remain pressed together by the movement (Fig. 5.5(c)). Heating continues even after the movement stops. The discharge continues from Figure 5.5(b) to (d). The timing chart of the prefusion method is shown in Figure 5.6. The moving distance shown in this chart is defined as the distance moved from the initial fiber endface position. The

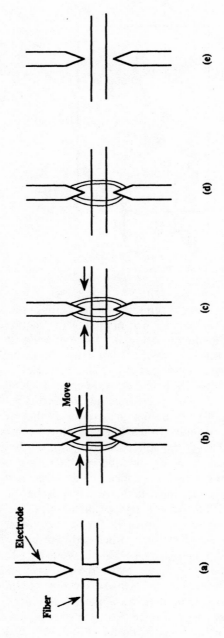

Figure 5.5 Prefusion method: (a) fiber set with a gap (imperfect fiber endface; (b) discharge starts (shaping of endface); (c) endfaces touch and are pressed together; (d) movement stops; (e) discharge stops (end of splice).

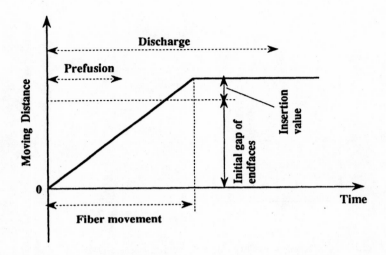

Figure 5.6 Timing chart of prefusion method.

prefusion discharge time is less than 1 second. The overall discharge time is a few seconds and the discharge time for splicing single-mode fibers is relatively short when compared to that for multimode fibers. Fibers of a few tens of microns are pressed and melted together. With this method, low-loss splices are obtained even for nonperfect endface quality. Since it is difficult to obtain good endface quality in the field, this method is beneficial for realizing a practical fusion-splicing machine.

5.3 HIGH-FREQUENCY DISCHARGE WITH HIGH-VOLTAGE TRIGGER METHOD

An electric discharge can be categorized as either a direct current (dc) discharge or an alternating current (ac) discharge. An ac discharge is preferred over a dc discharge. In the case of a dc discharge, only one electrode is worn out. On the other hand, two electrodes are symmetrically worn out for an ac discharge. The HHT method belongs with an ac discharge. As explained later in this section, a high-frequency discharge has good properties. We start here with the explanation of ac discharge properties.

5.3.1 AC Discharge Properties

This section presents experimental results for an ac discharge, and the results are explained by considering the movement of ions generated by a discharge. The relationship between discharge voltage and discharge current is shown in Figure 5.7.

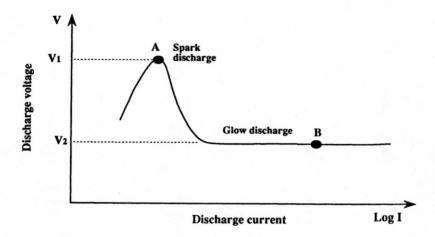

Figure 5.7 Relationship between discharge current and discharge voltage.

After a high voltage is applied to electrodes, an electric discharge takes place at the electrode gap. Before the spark discharge takes place at a high voltage V_1, a discharge current with a small value flows. Therefore, the current I in the horizontal axis is plotted in the logarithm scale (log I). After the spark discharge, more currents flow and the voltage drops to V_2. For example, V_2 is 500V and V_1 is 4 kV when the electrode gap $d = 1.5$ mm at a frequency of $f = 50$ Hz. The ac discharge properties are schematically shown in Figure 5.8, where the discharge voltage polarity is ignored [10]. The initial discharge is caused by the insulation breakdown in the air gap and it has a high voltage V_{si}. The air gap is partially ionized due to the initial spark voltage V_{si}. Therefore, the continuous discharge after the initial discharge has a lower spark voltage V_s. The spark voltage V_s is generated at every half cycle of the applied ac voltage. The V_{si} and V_g values in Figure 5.8 are nearly equal to V_1 and V_2 in Figure 5.7, respectively. The relationship among the voltages is expressed as follows:

$$V_{si} \geq V_s \geq V_g \tag{5.5}$$

These voltages depend on the gap between electrodes, and these dependencies, experimentally obtained, are shown in Figure 5.9. These voltages become higher along with the gap increase.

The initial spark voltage can be considered to be independent of the applied voltage frequency. The experimental relationships between the voltages V_g and V_s and the applied frequency are shown in Figures 5.10 and 5.11, respectively. Tungsten electrodes in a normal air atmosphere with a 1.5-mm gap length were used in the experiments. Both cone (angle $2\theta = 30$ deg) and wedge shapes were used for

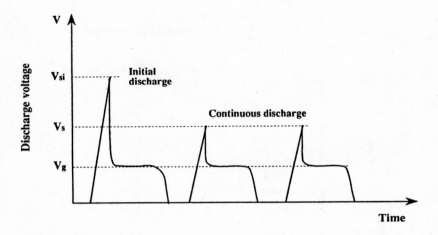

Figure 5.8 AC discharge of current and voltage (voltage polarity is ignored in this figure).

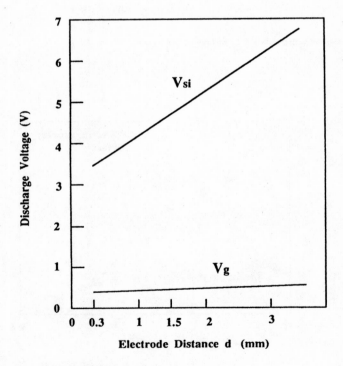

Figure 5.9 Electrode distance dependence of discharge voltages.

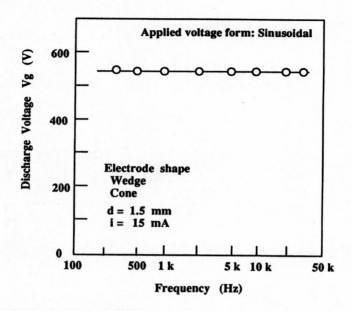

Figure 5.10 Frequency dependence of discharge voltage V_g. (After [8].)

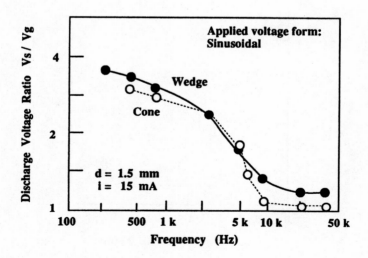

Figure 5.11 Frequency dependence of discharge voltage V_s. (After [8].)

the electrode geometry. It is found that V_g does not depend on the frequency. In Figure 5.11, the voltage V_s is normalized by V_g. In this case, the ratio V_{si}/V_g is about 8. It is found that the voltage V_s tends to approach the voltage V_g when the applied frequency becomes higher. The equation $V_s \simeq V_g$ holds when the frequency f is over 10 kHz. This is explained by the fact that ions created by a discharge cannot move along with the applied high voltage at high frequencies. Therefore, the created ions form a space charge between electrodes. The formed space charge results in dropping the V_s value. Therefore, the $V_s \simeq V_g$ condition is realized when the ion forms a space charge. A parameter T_0, defined by the following equation, is introduced for further discussion.

$$T_0 = \frac{d^2}{\mu V_g} \tag{5.6}$$

where d, μ, and V_g are the electrode gap, the ion mobility, and the discharge voltage defined in Figure 5.8, respectively. The parameter T_0 expresses the traveling time of ions from one electrode to another when the voltage V_g is applied to the electrodes. Equation (5.6) is derived from the model shown in Figure 5.12. The ion travels between electrodes with a speed of v. The traveling time T_0 is

$$T_0 = \frac{d}{v} \tag{5.7}$$

The speed v and the electric field E are expressed as

$$v = \mu E \tag{5.8}$$

$$E = \frac{V_g}{d} \tag{5.9}$$

By combining (5.7) through (5.9), (5.6) is obtained. When the polarity change of the applied voltage becomes faster than T_0, the ion flow cannot follow the applied voltage change and ions remain between the electrodes. These remaining ions form a space charge. To confirm this mechanism, the following experiments are made using a rectangular pulse for the applied voltage. The pulse waveform is shown in Figure 5.13. The transient time is defined by (5.10) for the parameter representing the waveform polarity change rate.

$$\hat{t} = t_r + t_d + t_f \tag{5.10}$$

where t_r is the rising time (the pulse voltage value: from 10% to 90%), t_d is the dead

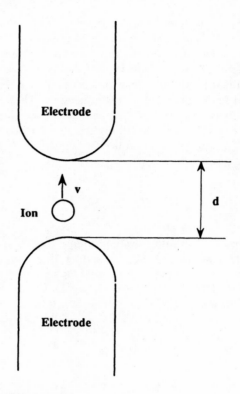

Figure 5.12 Model of the ion movement.

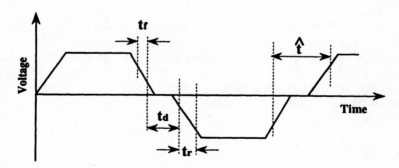

Figure 5.13 Definition of transient time.

time (from 10% to -10% through 0%), and t_f is the falling time (from 90% to 10%). The experimental results are shown in Figure 5.14 by varying $\hat{t}$ and d values at frequency $f = 1$ kHz. The discharge current is 15 mA and the electrode shape is a cone in this experiment. Both V_s and $\hat{t}$ are normalized by V_g and T_0, respectively. The T_0 value is calculated by (5.6), using the measured electrode gap d, the measured V_g, and $\mu = 2$ [cm^2/V · sec]. Figure 5.14 shows that V_s is nearly equal to V_g when $\hat{t} \lesssim T_0$ holds.

From the above experiments (Figs. 5.9 to 5.14), the following facts are found.

1. The ratio V_{si}/V_g is large.
2. The value V_g is independent of frequency f.
3. The equation $V_s \simeq V_g$ occurs in the following two cases: where the applied voltage frequency is higher than 10 kHz, and where the transient time for the applied voltage waveform is smaller than T_0, even for the low-frequency case.

When we use the higher frequency for the applied voltage, the transient time becomes smaller. Then the inequality $\hat{t} \lesssim T_0$ holds in the case of a high-frequency discharge.

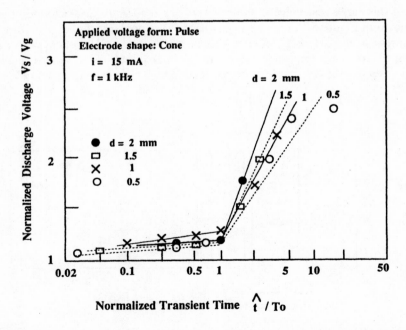

Figure 5.14 Relationship between the normalized V_s and the normalized transient time. (After [8].)

5.3.2 HHT Method

As the result of the above-mentioned fundamental experiment on an ac discharge, the HHT method was developed. First, the conventional method is explained. The power-supplying circuits used for the conventional discharge method use a transformer for voltage buildup from 100V to 4 kV with a commercial frequency ($f = 50$ Hz or 60 Hz). To realize the discharge characteristics shown in Figure 5.7 (voltage and current characteristics), the transformer in Figure 5.15 always generates a voltage higher than V_{si} (about 4 kV), and a resistance value R of a stabilized resistor is large (about 100 kΩ). After the initial discharge, the voltages V_s and V_g are imposed between electrodes. The remaining voltages ($V_{si} - V_s$) and ($V_{si} - V_g$) are consumed by this large resistance R. The electric power W necessary for a fusion splice is expressed as

$$W \approx V_g i = V_g (V_{si} - V_g)/R \tag{5.11}$$

where i is the current. The power consumption W_L by R is

$$W_L = Ri^2 = (V_{si} - V_g)^2/R \tag{5.12}$$

Although this large resistance R is necessary for realizing the discharge characteristics shown in Figure 5.7, the large resistance value prevents the reduction of power consumption in the discharge circuit. The large resistance puts a heavy load on the transformer, so that a large and heavy transformer is necessary. These facts result in large-sized and heavy fusion-splicing machines. The power ratio W_L/W is

$$W_L/W = (V_{si} - V_g)/V_g \tag{5.13}$$

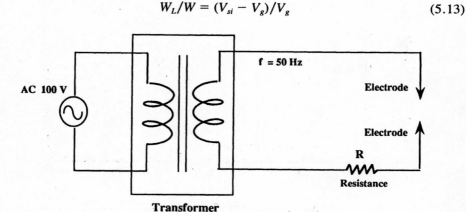

Figure 5.15 Power supply circuits for discharge in the conventional method.

For a 1.5-mm electrode gap, $V_{si} \approx 4,000\text{V}$ and $V_g \approx 550\text{V}$. Therefore, the ratio W_L/W is about 7. After the initial discharge, most of the power is consumed by the resistance R.

To reduce the power consumption, a high-voltage trigger method is adopted. The trigger voltage is higher than the initial spark voltage V_{si}. The trigger is applied only at the initial discharge, and relatively low voltage slightly higher than the spark voltage V_s is continuously applied. Therefore, the power ratio W_L/W for the trigger method is

$$W_L/W \approx (V_s - V_g)/V_g \qquad (5.14)$$

When the V_s value approaches the V_g value, the reduction in power consumption becomes more effective. Based on the experimental facts mentioned in the previous section, the following two methods are applicable for reducing power consumption with the trigger method. One method is to use the applied voltage with a frequency higher than 10 kHz. Another method is to use the low-frequency voltage, which is set to have the transient time $\hat{t}$ satisfy $\hat{t} \lesssim T_0$ (T_0 is the ion traveling time). For both methods, the transformer becomes lightweight because of the reduction in power consumption. In general, the transformer becomes lightweight when the frequency becomes higher. This is explained as follows. The transformer voltage generated by the magnetic field change is proportional to the frequency, and the transformer current does not depend on the frequency. Therefore, the actual transformer size decreases in the case of a higher frequency, when the allowable power is constant. Therefore, the reductions of both power consumption and weight are effectively accomplished when the HHT method is adopted. With this method, power consumption is reduced to one-seventh of that of the conventional method.

The discharge circuits for the HHT method are shown in Figure 5.16. The input dc voltage is 12V. A high frequency, ranging from 20 to 40 kHz, is obtained by

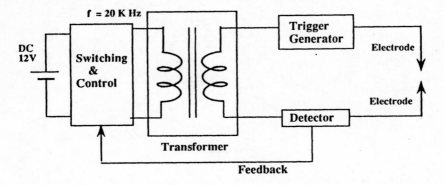

Figure 5.16 Power supply circuits for discharge in the HHT method.

switching the dc voltage (12V). Switching is done with semiconductor devices. This frequency range is selected because of the results in Figure 5.11 and because of the fact that these frequencies are popular in power supply circuits and many semiconductor devices are available. The trigger generator in Figure 5.16 is composed of diodes and capacitors. The feedback control is used in this circuit as shown in the figure.

It is necessary to investigate the effect of capacitance between electrodes and between one electrode and the surrounding metal, because the HHT method uses high frequencies. The electrode is approximated by a hyperboloid of revolution as in Figure 5.17 to investigate the effects analytically. Laplace's equation for a static field in the prolate spherical coordinates is solved by assuming that the static potential depends only on θ. The capacitance between electrodes is expressed in the following equation as a result of the solution:

$$C = \frac{10^{-9}}{36} \frac{d}{\cos \theta_0} \left[\log \frac{\cot (\theta_0/2)}{\cot ((\pi - \theta_0)/2)} \right]^{-1} \cdot [(2L/d + 1) \cos \theta_0 - 1] \quad (5.15)$$

where C, $2\theta_0$, L, and d are the capacitance between electrodes, the electrode angle, the electrode length, and the electrode gap, respectively. For example, the calculated

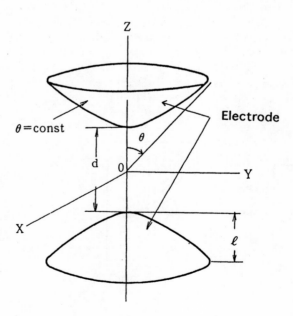

Figure 5.17 Model of electrodes.

capacitances are about 0.27 pF for $2\theta_0 = 30$ deg and $L = 20$ mm. The calculated capacitances between electrodes are found to be small for the parameters commonly used in actual fusion-splicing machines, and the effect of capacitance is negligible. This has also been confirmed experimentally [8].

5.3.3 Characteristics of High-Frequency Discharge

A space charge is formed between electrodes in the case of a high-frequency discharge. Therefore, this formed space charge affects the heat distribution of a high-frequency discharge, and it is anticipated to be different from that of a low-frequency discharge. The heat distribution difference between low- and high-frequency discharges is measured by using the setup shown in Figure 5.18 [18,10]. Electrodes used in the experiment are 1-mm-diameter tungsten rods with cone-shaped tips. An image of the light-emitting discharge area, enlarged by a microscope, is detected by a video camera. The light intensity distribution along a sample line on a TV monitor is observed as a video signal on an oscilloscope screen. When a sample line (a–a') is chosen so as to cross the electrode axis vertically, as shown in Figure 5.19(a), a light intensity distribution such as that in Figure 5.19(b) can be observed. This obtained light intensity is a summation of the emitted light from every point of a z-direction line, so that the emitting light strength distribution in the cross section of the discharge area, which is symmetrical with respect to the electrode axis, can be obtained as the Abel inversion of the observed distribution. Experiments were carried out for discharges of $f = 50$ Hz and $f = 20$ kHz for a 1.5-mm electrode gap condition. The Abel inversion was performed for the observed light distribution profiles for several sample lines vertical to the electrode axis. The observed distributions were almost Gaussian profiles, so that the inverse distribution profiles along the

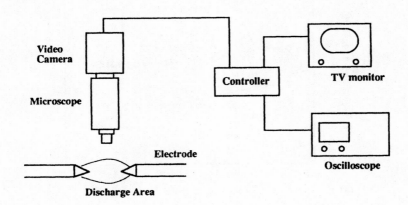

Figure 5.18 Setup for measurement. (After [18].)

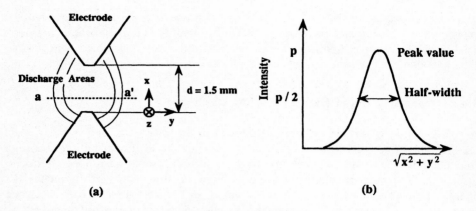

(a)   **(b)**

Figure 5.19 Measurement for heat distribution. (After [18].)

radial coordinate, which originate from the electrode axis, also became nearly Gaussian. The observed results for spatial distributions, peak values and half widths, are shown in Figure 5.20(a,b), respectively [18,10]. From these results, it can be found that the 20-kHz discharge has smaller peak values and larger half-width values than the 50-Hz discharge. This means that the 20-kHz discharge gives flatter spatial distribution of discharge energy, which is desirable for fiber mass-fusion splicing (this will be explained later). The presence of ions (space charge) makes the discharge energy distribution stable. These experiments indicate that the high-frequency discharge has a more stable and uniform heat distribution than the low-frequency discharge.

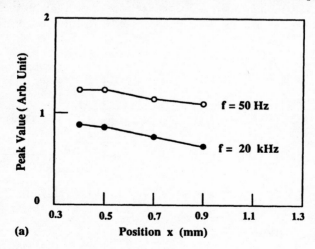

(a)

Figure 5.20 Measured results for spatial distribution of discharge. (After [18,10].)

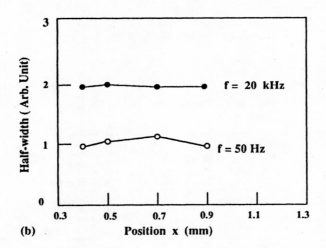

(b)

Figure 5.20 (Continued)

5.4 FIBER-ALIGNMENT METHOD

Both prefusion and stable discharge methods are important for obtaining low-loss splices, as explained in the previous sections. Along with these methods, fiber alignment is also important for obtaining low-loss splices. Several fiber-alignment methods have been proposed and used, and they are listed in Table 5.3. They are divided into fixed and nonfixed methods.

Fibers are not moved laterally in the fixed method; they are only placed in the predetermined positions, such as V-grooves. This method is also called passive alignment, and is widely used in practical fusion-splicing machines because of its simple mechanism. When fibers with a small core eccentricity are used, this method is applicable not only for multimode fibers but also for single-mode fibers. Since fiber manufacturing process technologies have improved, today's fibers are accurately produced. Therefore, mass-fusion splices of single-mode fibers with a splice loss lower than 0.05 dB have been easily obtained in the field by a splicing machine using the fixed method.

In the nonfixed method, fibers are moved laterally to obtain accurate positions before the discharge. This method is also called active alignment. Many nonfixed methods have been proposed, especially in the early stages of the development of the single-mode fiber splice. At that time, fibers had a relatively large core eccentricity compared to today's fibers. Transmitted optical power is measured and the measured value is used for alignment in the power monitoring methods. The power monitoring methods, shown in Figure 5.21, are divided into three categories: (1)

Table 5.3
Classification of Fiber-Alignment Method

Fixed or Not	Method		Application
Fixed (passive alignment)	Mechanical positioning method (e.g., V-groove alignment)		A, C
Movable (active alignment)	Power monitoring method	Three points	B
		Two points	
		One point (LID)	
	Visual method	Outer diameter	A
		Core (direct core monitoring)	B
	Light sensor	Outer diameter	A

A: Splice of multimode fibers, splice of single-mode fibers with small core eccentricity; B: splice of single-mode fibers with relatively large core eccentricity, splice with very small splice loss; C: mass-fusion splice.

three-points method, (2) two-points method, and (3) one-point method. In the three-points method, a light source, a fusion splicer, and a detector must be used, and they are placed in three different locations (three points) in the field as shown in Figure 5.21(a). To make use of the detected optical power for fiber alignment as a feedback signal, something to transmit the detected signals is required. The results of a field trial using the three-points method at an early investigation stage for single-mode fiber splices were reported in [19], and the obtained mean splice loss was about 0.1 dB with a standard deviation of 0.08 dB. A twisted pair (metallic pair wire) was used to transmit the detected signals in this field trial. In the two-points method, a detector is placed near or in a fusion splicer (Fig. 5.21(b)). Therefore, equipment is placed in only two locations. The transmission power is locally detected (local detection). One typical method for local detection is by bending a fiber so that the radiated light caused by bending is detected. In the one-point method, a light source, a fusion splicer, and a detector are all placed in one location (Fig. 5.21(c)). This method is also called the local injection and detection (LID) method. One example of LID has been reported in [20]. Local injection and local detection are realized by bending a fiber in most cases. In both the two-points and one-point methods, a means of transmission, such as metallic pair cables, is unnecessary.

Visual methods are other alignment methods that belong to the nonfixed methods. In one of the visual methods, the outer diameters of both mated fibers are viewed through a microscope in one direction or in two perpendicular directions. For viewing the two perpendicular directions, a mirror is placed near fibers, as shown

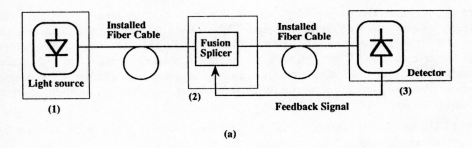

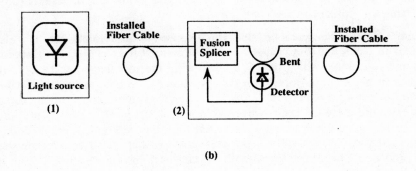

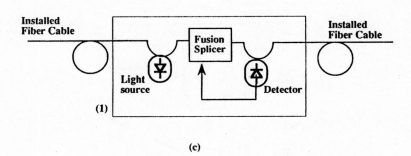

Figure 5.21 Power monitoring methods: (a) three-points; (b) two-points (local detection); (c) one-point (LID).

in Figure 5.22. The magnified picture is directly viewed by an operator or is indirectly viewed with a TV monitor using a TV camera. By using the visual information of the outer-diameter misalignment, fibers are aligned according to the outer-diameter reference. This method is powerless for a single-mode fiber with a core eccentricity. However, direct core monitoring is a powerful method for a single-mode fiber with a core eccentricity. Several direct core monitoring methods have been proposed:

1. Using the fluorescence of a germanium-doped silica core excited by ultraviolet light [21];
2. Using a differential interference contrast microscope [22];
3. Using a beam splitter and local injected lights from both directions [23];
4. Using an ordinary microscope [24–27].

In the method using fluorescence, fibers are exposed to ultraviolet light. The electron transitions in GeO_2 and GeO caused by the ultraviolet light exhibit the fluorescence in a germanium-doped silica core [28]. By using the fact that only the core region fluoresces in germanium-doped core fibers, direct core monitoring is possible. In [21], core monitoring is carried out by the magnified fluorescence image through a microscope, and a He-Cd laser (wavelength $\lambda = 0.325$ μm) is used as a ultraviolet light. The second method uses a differential interference contrast microscope. The third method uses a beam splitter, which is inserted between fibers before the discharge, and lights from a tungsten halogen lamp are locally injected in both direc-

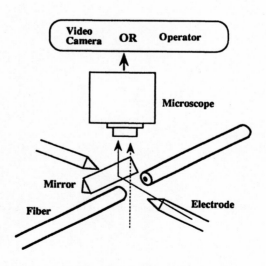

Figure 5.22 Visual alignment method in two perpendicular directions (outer diameter).

tions by bending both fibers. The purpose of the local injection light is to be able to observe both cores clearly through a microscope. The fourth method uses an ordinary microscope and the setup is shown in Figure 5.23. The core detection conditions are discussed in [26,27], which use the ordinary microscope method, and it has been shown that to obtain a clear core image, a microscope with both a small depth of focus and a high resolution is required. It has also been shown that there is an optimum location for the objective plane in order to minimize an error of alignment with this method. Alignment errors are made when determined from the visual image, so that the alignment is corrected [24]. Visual methods are very attractive, because operators and equipment are located at one point. Among the nonfixed alignment methods (the active alignment methods), the method using the ordinary microscope is the most widely used in today's commercial fusion-splicing machines.

The alignment method using a light sensor also belongs to the nonfixed alignment category. In the proposed method of [29], fiber positions in two perpendicular

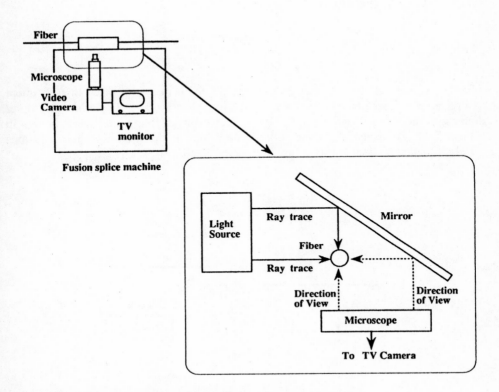

Figure 5.23 Direct core monitoring method (using an ordinary microscope).

directions are detected by two light sensors for each fiber, so four sensors are used in the alignment of two mated fibers.

In the case of the nonfixed alignment methods, fine movement mechanisms for precise alignment are required. Several mechanisms have been proposed and discussed in [30]. These proposed mechanisms are the precision movable stage driven by a motor, the piezoelectric device, and the elastic deformation device. Required characteristics such as linearity of movement, maximum displacement, and size are different for both alignment methods and fusion-splicing machine design.

5.5 SURFACE TENSION EFFECT

The surface tension of silica glass fibers melted by discharge heat is generated in a fusion-splice process. This surface tension tends to align two outer surfaces of mated fibers during heating, and this is known as the surface tension effect [7]. A model of fibers during heating is shown in Figure 5.24, and the force W is expressed as [7]

$$W \approx \frac{2\pi bT}{s} x \qquad (5.16)$$

where T, x, and s are the surface tension of melted silica glass, the axis misalignment of the outer surface, and the melted width, respectively, and $2b$ is the fiber outer diameter. The force W tends to recover the outer-surface alignment, as indicated in Figure 5.24. By considering the force equilibrium at the melted region, the axis misalignment x is obtained as a function of the heating time t [7]:

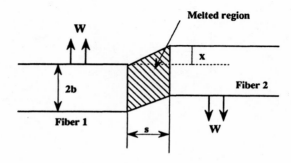

Figure 5.24 Surface tension effect. (After [7].)

$$x = x_0 \exp\left(-\frac{2T}{b\mu} t\right) \tag{5.17}$$

where μ is the viscosity of melted silica glass and x_0 is the initial misalignment.

The surface tension effect is beneficial or harmful, depending on the situation. When fibers with small core eccentricity are used, the fixed method, such as a fixed V-groove, is applicable not only for multimode fibers but also for single-mode fibers. In this case, the surface tension effect is beneficial for reducing the misalignment in the butt-joint state (before the discharge state), and the effect serves as a self-alignment (self-alignment effect [7]). However, when fibers with large core eccentricity are used and a nonfixed alignment method is adopted, the surface tension effect is harmful [30,31]. The surface tension effect tends to reduce axis misalignment concerning the outer surface, not the core position. Therefore, the core alignment achieved by the nonfixed alignment method is destroyed by the surface tension effect. To prevent this effect in the core alignment of single-mode fibers with the nonfixed method, a relatively short discharge time is adopted compared to the discharge time for multimode fiber splices using the fixed method [30,31].

To prevent the surface tension effect in the core alignment fibers with the non-fixed method, an method that intentionally shifts the axis has been proposed [32]. In this method, the intentional misalignment (over-compensated alignment) is added to core alignment before discharge. The added misalignment ΔX is canceled by the axis movement Δx during the heating time, when $\Delta X = \Delta x$. By using this method, very low splice losses have been obtained even for fibers with a large core eccentricity.

5.6 UNIFORM HEATING FOR MASS-FUSION SPLICE

Mass-fusion splicing is a fusion-splicing method that splices multiple fibers with one discharge. That is, fibers set in a row (such as fibers in a fiber ribbon) are spliced with simultaneous heating. How to realize uniform heating is the key to the mass-fusion-splicing method.

5.6.1 Uniform Heating With Two Pairs of Electrodes

Although uniform heating for five fibers in a row is obtained by using one pair of electrodes in the case of a low-frequency discharge ($f = 50$ Hz), the discharge has been found to be unstable [9,33]. By using two pairs of electrodes, uniform heating has been realized in the case of a low-frequency discharge [9,33]. The discharge heating method using two pairs of electrodes is shown in Figure 5.25. Diodes are inserted between electrodes and a high-voltage source with $f = 50$ Hz. The discharges take place between electrode A and A' and between B and B' alternately.

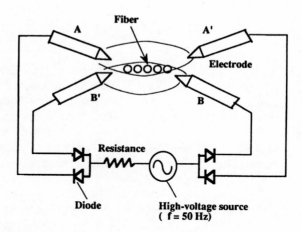

Figure 5.25 Discharge heating method using two pairs of electrodes. (After [9].)

By optimizing the electrode and fiber positions, uniform heating (temperatures of fibers are about 2,000°C and are controlled to within ±25°C) has been realized [9]. However, when used in the field, this method has the disadvantage that four electrodes must be replaced in the case of electrode consumption, and it is inconvenient when compared to a pair of electrodes (two electrodes), because the four electrodes must be accurately positioned.

5.6.2 Uniform Heating Using High-Frequency Discharge

A high-frequency discharge has a broader and more stable heating region when compared to a low-frequency discharge, as discussed before. Here, a high-frequency discharge is discussed for uniform heating. When set on the electrode axis, the discharge energy between a pair of electrodes is distributed unevenly, so that fibers in a row cannot be heated up to the same temperature. To achieve uniform heating, the fibers are offset from the electrode axis (Fig. 5.26). The fiber temperature variance diminishes when the fibers are suitably offset ($x \simeq 0.2$ mm) for a high-frequency discharge ($f = 20$ kHz). In the case of a low-frequency (50-Hz) discharge, there is also a suitable offset ($x \simeq 0.5$ mm), but the fiber temperature is unsteady. For this reason, a 50-Hz discharge is not applicable to mass-fusion splicing when the two-pair electrode technique is not used. The reason for the heating condition

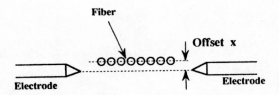

Figure 5.26 Heating method using a high-frequency discharge and a pair of electrodes. (After [10].)

stability of the high-frequency discharge for mass-fusion splicing is schematically shown in Figure 5.27, where the curves indicate the difference of the discharge energy distribution and of the suitable fiber offset from the electrode axis. These curves are drawn based on the experimental results shown in Figure 5.20. For a high-frequency discharge, fibers are set closer to the electrode axis, and discharge energy is distributed more moderately, so the gradient of the discharge energy at the fiber position is smaller than in the case of a low-frequency discharge. Therefore, in a high-frequency discharge, fiber temperatures are more stable against the positional disturbance (variation of offset x).

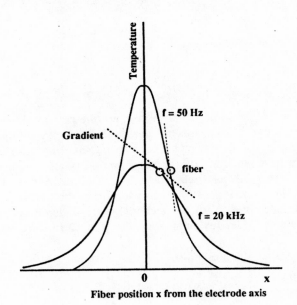

Figure 5.27 Comparison of heating condition for two types of discharge. (After [10].)

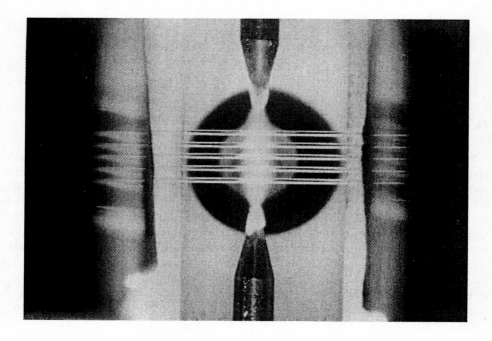

Figure 5.28 Five fibers being fusion-spliced.

In addition to the advantage that more than one pair of electrodes are not required for the discharge heating mechanism, high-frequency discharge heating has
another advantage in that its power supply can be small because of voltage transformer efficiency improvement. Power consumption is also reduced by the application of the HHT method, as discussed before. A photograph of five fibers being
fusion-spliced is shown in Figure 5.28. In addition to the uniform-heating technology, mass-cutting and fiber-alignment technologies are also important for realizing
the mass-fusion splice. Mass-fusion splicing of up to 12 fibers has been realized so
far.

5.7 PROTECTION

Generally, the coating is removed when coated fibers are spliced. Throughout the
splicing process, that is, coating removal, fiber cutting, placing on fusion-splicing
machine, and even in the heating process, cracks are introduced, and they weaken

the fiber strength. Typical measured fiber strength for a fiber with a 125-μm diameter is shown in Figure 5.29. The strength of fusion-spliced fibers decreases to be about 10% of the original coated fibers. Therefore, there must be protection (reinforcement) for the spliced part. Several protection methods have been proposed and used, and some of them are listed in Table 5.4 [34,35,10]. Chemical reaction, hot-melt, and UV-curable resin have been proposed as adhesives. Among these, the hot-melt type, which is cured by heat, has been widely used. For the protection structure, a V-groove, a tube, and sandwich plates have been proposed. In the injection mold method, spliced fibers are placed in a mold and plastic is injected. The strength of fibers after protection recovers, as indicated in Figure 5.29. Several factors must be considered for selecting the protection methods: (1) reliability (splice loss variation and break), (2) good workability, and (3) cost. When the protection method and its design are not good, splice loss varies a great deal with temperature change. With bad protection, splice loss increases, and finally spliced fibers break. With a suitable design and method, splice loss variation is very small (e.g., within 0.02 dB from $-30°$ to $+60°$C change) and long-term reliability is ensured.

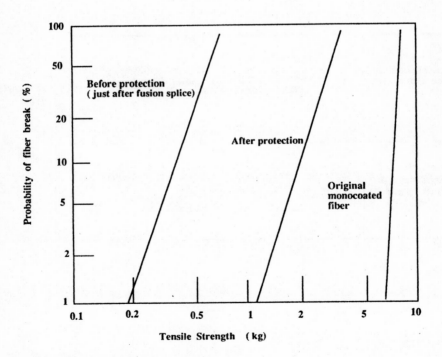

Figure 5.29 Strength of fibers with a 125-μm diameter.

Table 5.4
Protection Methods

Method	Structure	Adhesive
Plastic V-groove with a cover	Cover · Spliced Fiber · Plastic V-groove	Chemical reaction type
Injection mold	Spliced Fiber · Plastic mold	Hot-melt type
Heat shrinkable tube with a steel rod	Heat shrinkable tube · Spliced Fiber · Steel rod	
Sandwich using glass ceramic plates	glass ceramic plate · Spliced Fiber	

5.8 FUSION-SPLICING MACHINE

The automatic fusion-splicing machines for multimode and single-mode fibers have been developed and commercially available using the prefusion and HHT methods. They can be operated both with a small battery (dc 12V) and with a commercial power supply. As an example, a photograph of a fusion-splicing machine, developed in 1980, is shown in Figure 5.30. At first, a fusion-splicing machine using the three-points or two-points method for fiber alignment was developed in order to splice single-mode fibers. Later on, a fusion-splicing machine using the direct core monitoring method or the LID method for alignment was developed for single-mode fibers. A mass-fusion-splicing machine for a fiber ribbon, which contains up to 12 fibers, was developed using the uniform-heating method with a pair of electrodes (high-frequency discharge). The direct core monitoring splice machine and mass-fusion-splicing machine adopt the prefusion and HHT methods. A fully automatic fusion-splicing machine was also investigated and developed [36]. All these fusion-splicing machines, including a mass-fusion-splicing machine and a fully automatic fusion-splicing machine, are commercially available at present.

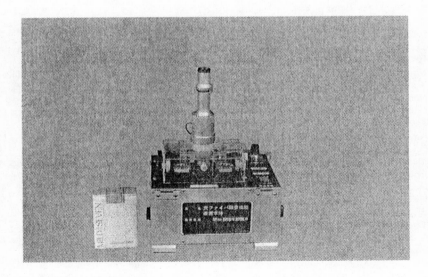

Figure 5.30 A fusion-splicing machine.

REFERENCES

[1] Bisbee, D. L., "Optical Fiber Joining Technique," Bell Syst. Tech. J., Vol. 50, 1971, p. 3153.

[2] Fujita, H., Y. Suzaki, and A. Tachibana, "A Method of Splicing Optical Fibers," IEICE of Japan, Technical Report OQE75–63, 1975 (in Japanese); also, H. Fujita, Y. Suzaki, and A. Tachibana, "Optical Fiber Splicing Technique With a CO2 Laser," Appl. Opt., Vol. 15, 1976, p. 320.

[3] Hatakeyama, I., and H. Tsuchiya, "Fusion Splices for Optical Fibers," IECE of Japan, Technical Report OQE75–92, 1975 (in Japanese); also, I. Hatakeyama and H. Tsuchiya, "Fusion Splices for Optical Fibers by Discharge Heating," Appl. Opt., Vol. 17, 1978, p. 1959.

[4] Kohanzadeh, Y., "Hot Splices of Optical Waveguide Fibers," Appl. Opt., Vol. 15, 1976, p. 793.

[5] Jocteur, R., and A. Tarday, "Optical Fibers Splicing With Plasma Torch and Oxhydric Micro-burner," European Conf. on Optical Communications (ECOC'76), 1976.

[6] Hirai, M., and N. Uchida, "Melt Splice of Multimode Optical Fiber With an Electric Arc," Electron. Lett., Vol. 13, 1977, p. 123.

[7] Hatakeyama, I., and H. Tsuchiya, "Fusion Splices for Single-Mode Optical Fibers," IEEE J. Quantum Electron., Vol. QE-14, 1978, p. 614.

[8] Kashima, N., and F. Nihei, U.S. Pat. 4,383,844; also, N. Kashima and F. Nihei, "Optical Fiber Fusion Splice Using High Frequency Discharge With High Voltage Trigger," IECE of Japan, Vol. E64, 1981, p. 529.

[9] Tachikura, M., "Fusion Mass-Splicing for Optical Fibers by Discharge Heating," Electron. Lett., Vol. 17, 1981, p. 694.

[10] Tachikura, M., and N. Kashima, "Fusion Mass-Splices for Optical Fibers Using High-Frequency Discharge," IEEE J. Lightwave Technol., Vol. LT-2, 1984, p. 25.

[11] Gloge, D., P. W. Smith, D. L. Bisbee, and E. L. Chinnock, "Optical Fiber End Preparation for Low-Loss Splices," Bell Syst. Tech. J., Vol. 52, 1973, p. 1579.

[12] Caspers, F. R., and E. G. Neumann, "Optical-Fiber End Preparation by Spark Erosion," Electron. Lett., Vol. 12, 1976, p. 443.

[13] Chesler, R. B., and F. W. Dabby, "Fibers Simple Testing Methods Give Users a Feel for Cable Parameters," Electronics, 5 August 1976, p. 90.

[14] Khoe, G. D., and G. Kuyt, "Cutting Optical Fibers With a Hot Wire," Electron. Lett., Vol. 13, 1977, p. 147.

[15] Kinoshita, K., and M. Kobayashi, "End Preparation of Fusion Splicing of an Optical Fiber Array With a CO2 Laser," Appl. Opt., Vol. 18, 1979, p. 3256.

[16] Holloway, D. G., "The Physical Properties of Glass," Wykeham Publications, 1973.

[17] Haibara, T., M. Matsumoto, and M. Miyauchi, "Design and Development of an Automatic Cutting Tool for Optical Fibers," IEEE J. Lightwave Technol., Vol. LT-4, 1986, p. 1434.

[18] Kashima, N., M. Tachikura, and F. Nihei, "Heat Distribution Measurement of Discharge in Optical Fiber Fusion Splice," IECE of Japan, Vol. J65-C, 1982, p. 721 (in Japanese).

[19] Tanifuji, T., Y. Kato, and S. Seikai, "Realization of a Low-Loss Splice for Single-Mode Fibers in the Field Using an Automatic Arc-Fusion Splicing Machine," Conf. on Optical Fiber Communication (OFC'83), MG3, 1983.

[20] De Block, C. M., and P. Matthijsse, "Core Alignment Procedure for Single-Mode-Fiber Joining," Electron. Lett., Vol. 20, 1984, p. 109.

[21] Tatekura, K., H. Yamamoto, and M. Nunokawa, "Novel Core Alignment Method for Low-Loss Splicing of Single-Mode Fibers Utilizing UV-Excited Fluorescence of Ge-Doped Silica Core," Electron. Lett., Vol. 18, 1982, p. 712.

[22] Haibara, T., M. Matsumoto, T. Tanifuji, and M. Tokuda, "Monitoring Method for Axis Alignment of Single-Mode Optical Fiber and Splice-Loss Estimation," Opt. Lett., Vol. 8, 1983, p. 235.

[23] Imon, K., and M. Tokuda, "Axis-Alignment Method for Arc-Fusion Splice of Single-Mode Fiber Using a Beam Splitter," Opt. Lett., Vol. 8, 1983, p. 502.

[24] Kawata, O., K. Hoshino, and K. Ishihara, "Low-Loss Single-Mode-Fiber Splicing Technique Using Core Direct Monitoring," Electron. Lett., Vol. 19, 1983, p. 1048.

[25] Kitazawa, I., S. Nishi, "Single-Mode-Fiber Core Alignment by the Focusing Method," Conf. on Optical Fiber Communication (OFC'84), TUM6, 1984.

[26] Katagiri, T., and M. Tachikura, "Consideration on Core Detecting Conditions of Single-Mode Optical Fiber," IECE of Japan, Vol. J66-B, 1983, p. 1520 (in Japanese).

[27] Katagiri, T., M. Tachikura, and I. Sankawa, "Optical Microscope Observation Method of a Single-Mode Optical-Fiber Core for Precise Core-Axis Alignment," IEEE J. Lightwave Technol., Vol. LT-2, 1984, p. 277.

[28] Yuen, M. J., "Ultraviolet Absorption Studies of Germanium Silicate Glasses," Appl. Opt., Vol. 21, 1982, p. 136.

[29] Turday, A., M. Jurczyczyn, L. Jeunhomme, M. Carratt, and R. Hakoun, "Automatic Single Mode Fiber Splicing Machine Without Power Monitoring," 9th European Conf. on Optical Communications (ECOC), 1983, p. 113.

[30] Toda, Y., O. Watanabe, M. Ogai, and S. Seikai, "Low-Loss Fusion Splice of Single Mode Fiber," International Communication Conf. (ICC'81), No. 27.7, 1981.

[31] Kato, Y., S. Seikai, and M. Tateda, "Arc-Fusion Splicing of Single-Mode Fibers 1: Optimum Splice Conditions," Appl. Opt., Vol. 21, 1982, p. 1332.

[32] Kawata, O., K. Hoshino, and K. Ishihara, "Core-Axis-Alignment Method to Achieve Ultra-Low-Loss Fusion Splicing for Single-Mode Optical Fibers," Opt. Lett., Vol. 9, 1984, p. 255.

[33] Tachikura, M., "Fusion Mass-Splicing for Optical Fibers Using Electric Discharges Between Two Pairs of Electrodes," Appl. Opt., Vol. 23, 1984, p. 492.

[34] Hirai, M., S. Seikai, N. Kashima, M. Shimoda, and N. Uchida, "Arc-Fusion Splice and Splice Machine for Multimode Fibers—Pre-Fusion Method," Review of ECL, Vol. 27, 1978, p. 2467 (in Japanese).

[35] Miyauchi, M., M. Matsumoto, Y. Toda, K. Matsuno, and Y. Tokumaru, "New Reinforcement for Arc-Fusion Spliced Fiber," Electron. Lett., Vol. 17, 1981, p. 907.

[36] Arioka, R., T. Haibara, and M. Tachikura, "Fully-Automatic Optical Fiber Splice Machine," International Wire and Cable Symp. (IWCS), 1984, p. 50.

Chapter 6
Fusion Splice Using Gas Laser and Flame

The method of fusion splicing with an electric discharge has been widely used. In principle, heating sources other than the electric discharge, which can melt fibers, are possible for fusion splices. In this chapter, the fusion-splicing method using a gas-laser and a flame are explained. These methods have been proposed for splicing silica fibers. Fusion splicing with a CO_2 gas laser has been investigated not only for a single fiber, but also for a fiber ribbon. Fusion splicing with a flame of a microburner has been investigated for a single fiber.

6.1 FUSION SPLICING USING GAS LASER

A CO_2 laser is used for welding several materials in industry, and this laser is also applied to fiber fusion splicing [1,2]. A laser beam is focused on fibers with a lens. The required laser power is calculated based on the following thermal equation based on a 1D model [1].

$$\frac{\partial T}{\partial t} = \frac{K}{\rho C_p} \frac{\partial^2 T}{\partial x^2} - \frac{4hT}{\rho C_p d} \tag{6.1}$$

where T, h, ρ, C_p, d, and t are the temperature, thermal conductivity, heat transfer coefficient, density, specific heat, diameter of a fiber, and time, respectively. The heat flux Q_0 from a heat source is assumed to obtain temperature T by solving (6.1). By setting the obtained temperature $T = T_s$ (the melting point), the required laser power for melting silica fibers is obtained as [1]

$$Q_0 = 2T_s \left(\sqrt{Kh/d}\right) \left[\text{erf}\left(\sqrt{(4h/\rho C_p d)}\,t\right)\right]^{-1} \tag{6.2}$$

When the values for silica fibers are $K = 3.5 \times 10^{-3}$ cal/s cm°C, $h = 1 \times 10^{-2}$ cal/s cm^2°C, $T_s = 2,000$°C, $C_p = 0.25$ cal/g°C, $\rho = 2.2$ g/cm^3, and $d = 1.5 \times$

10^{-2} cm, the required Q_0 is about 810 W/cm^2 for 1 to 5 seconds of heating time [1]. This value is obtained from the point source model. For a fiber with a 125-μm diameter (150-μm diameter), the required laser power is about $0.1W$ ($0.15W$) based on the assumption that the laser beam ω is equal to the fiber diameter d. The actual laser beam has a Gaussian profile; therefore, it is considered to be a Gaussian heat source. More accurate analysis based on a Gaussian heat source is made in [3], and the required laser power is slightly higher than that of the point source. Experiments reveal that the laser power is $0.5W$ to $3W$ for a fiber with a 150-μm diameter [1–3]. The actual laser beam used for a splice is usually set to be wider than a fiber diameter. This may be the main reason for the discrepancy between the calculation and the experiment. Other factors of the discrepancy are the reflection of laser light at the fiber surface, the loss of laser power due to a focusing lens absorption, and light scattering.

Splices of 1D- and 2D-arrayed fibers have been demonstrated [4,5]. The setup is shown in Figure 6.1. Fibers are set in a V-groove stage, which is not shown in the figure. Fibers are spliced one by one by moving the stage in the case of 1D-arrayed fibers (Fig. 6.1(a)). Therefore, this is different from mass-fusion splicing with an electric discharge, where 1D-arrayed fibers (fibers in a fiber ribbon) are spliced simultaneously with one discharge. In the case of 2D-arrayed fibers, fibers are aligned using a V-groove stage of a layered structure. For fusion splicing one

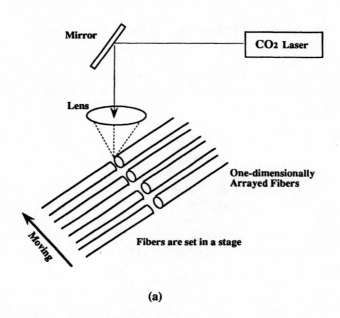

(a)

Figure 6.1 Fusion splicing using CO_2 laser. (After [4,5].)

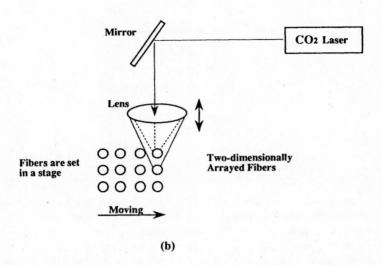

(b)

Figure 6.1 (Continued)

by one, both a focusing lens (germanium lens) and a stage are moved to focus the laser beam on the target fibers to be spliced (Fig. 6.1(b)). Fibers in an array are spliced by the movement of the stage, and fibers in a layered position are spliced by the movement of the lens. This method is also basically the one-by-one splicing method. Splicing of 8×3 fibers (3-layer 8-fiber array: 24 fibers) has been demonstrated [5].

The application of fusion splicing with a CO_2 laser to single-mode fibers has been demonstrated [6]. The principle is the same as the one described above. In [6], paraboloidal and spherical mirrors are used instead of a lens for focusing the laser beam. The laser power has been reported to be about $0.6W$ in this case.

6.2 FUSION SPLICING USING A FLAME

A flame from a microburner used as a heat source for fusion splicing has been demonstrated using an oxyhydrogen torch [7]. The method is a straightforward extension of the heating method used in silica glass manufacturing. Two mated fibers are aligned using V-grooves, and a microburner is positioned below the fibers (Fig. 6.2). The flame heating time has been reported to be 20 to 30 seconds or longer [8]. The flame heating time depends on the flame temperature, and low-loss flame fusion splices can be obtained over a wide heating-time range (e.g., from 1 to 150 seconds) [9]. The fiber heating region obtained with the flame method is broad compared to that of the electric-discharge method. This characteristic may be useful for applying the

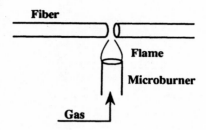

Figure 6.2 Splicing with a flame.

flame heating method to manufacturing fiber-type passive optical components. The flame fusion-splicing method is applied to obtain a very strong fusion splice. Generally speaking, we must take care to prevent fiber surface cracks made by contacts for obtaining strong fusion splices. Cracks are generated by removing the coating, cutting the fiber, and the process of setting the fiber in the case of the electric-discharge fusion-splicing method (Fig. 5.1). The process is similar for gas laser and flame fusion-splicing methods. Very strong fusion splices have been obtained by replacing the ordinary oxyhydrogen flame with a chlorine-hydrogen flame with the aid of oxygen [10,11]. By using the chlorine-hydrogen flame, the tensile strength of fusion splices is reported to be double that using the ordinary oxyhydrogen flame [10]. According to [10], this is due to the fact that OH corrosion degrades the strength of the splice during heating, and chlorine replaces the hydroxyl ions on a silica surface. By optimizing the flame heating temperature and the flame composition, very strong fusion splices with the original fiber strength were obtained [11].

Generally speaking, the application of the flame fusion-splicing method may be restricted to special cases, because the handling of a flame is inconvenient.

REFERENCES

[1] Fujita, H., Y. Suzaki, and A. Tachibana, "A Method of Splicing Optical Fibers," IECE of Japan, Technical Report OQE75–63, 1975 (in Japanese); also, H. Fujita, Y. Suzaki, and A. Tachibana, "Optical Fiber Splicing Technique With a CO_2 Laser," Appl. Opt., Vol. 15, 1976, p. 320.

[2] Egashira, K., and M. Kobayashi, "Optical Fiber Splicing With a Low-Power CO_2 Laser," Appl. Opt., Vol. 16, 1977, p. 1636.

[3] Egashira, K., and M. Kobayashi, "Analysis of Thermal Conditions in CO_2 Laser Splicing of Optical Fibers," Appl. Opt., Vol. 16, 1977, p. 2743.

[4] Kinoshita, K., and M. Kobayashi, "End Preparation and Fusion Splicing of an Optical Fiber Array With a CO_2 Laser," Appl. Opt., Vol. 18, 1979, p. 3256.

[5] Kinoshita, K., and M. Kobayashi, "Two-Dimensionally Arrayed Optical-Fiber Splicing With a CO_2 Laser," Appl. Opt., Vol. 21, 1982, p. 3419.

[6] Rivoallan, L., J. Y. Guilloux, and P. Lamouler, "Monomode Fiber Fusion Splicing With CO_2 Laser," Electron. Lett., Vol. 19, 1983, p. 54.

[7] Jocteur, R., and A. Tarday, "Optical Fibers Splicing With Plasma Torch and Oxhydric Micro-burner," European Conf. on Optical Communications (ECOC'76), 1976, p. 261.

[8] Rondan, B. F., and A. A. Morales, "Optical and Mechanical Characteristics of Microflame Fusion Splicing of Optical Fiber," International Wire and Cable Symp. (IWCS), 1982, p. 163.

[9] Krause, J. T., W. A. Reed, and K. L. Walker, "Splice Loss of Single-Mode Fiber as Related to Fusion Time, Temperature, and Index Profile Alteration," IEEE J. Lightwave Technol., Vol. LT-4, 1986, p. 837.

[10] Krause, J. T., C. R. Kurkjian, and U. C. Paek, "Tensile Strength > 4 GPa for Lightguide Fusion Splices," Electron. Lett., Vol. 17, 1981, p. 812.

[11] Krause, J. T., and C. R. Kurkjian, "Fiber Splices With 'Perfect Fiber' Strengths of 5.5 GPa, $\nu < 0.01$," Electron. Lett., Vol. 21, 1985, p. 533.

Chapter 7

Mechanical and Adhesive-Bonded Splices

Mechanical and adhesive-bonded splices are explained in this chapter. Neither method uses heat, in contrast with the fusion-splicing methods such as electric discharge, gas laser, and flame fusion splicing. Therefore, both mechanical and adhesive-bonded splices have several things in common with regard to fiber positioning. The common factors and differences are explained first. Then both methods are explained by using examples.

7.1 OUTLINE OF MECHANICAL AND ADHESIVE-BONDED SPLICES

Similar fiber-positioning methods are used for both mechanical and adhesive-bonded splicing methods. The basic alignment method uses V-grooves made from hard or soft materials, as shown in Figure 7.1.

When soft materials are used for a V-groove, it is possible to splice fibers with different diameters. The soft material can be deformed so that the centers of two fibers coincide (Fig. 7.1(b)). The usual alignment is the fixed (passive alignment) method. The power monitoring method (one of the nonfixed, or active alignment, methods) is used in special cases. Since the spliced fibers are not fused together in both methods, fiber positioning relies on an accurate substrate, such as a V-grooved substrate, not only at the splice time but also after the splice time. The key technologies for both methods are to realize an accurate substrate with low lost and ensure long-term reliability. There are many materials proposed and used for the substrates, such as silicon, metal, plastics, and steel. When the thermal expansion coefficient of the substrate is similar to that of the silica fibers, it is easy to suppress the loss variation due to temperature change and ensure long-term reliability. The values of the thermal expansion coefficient for several materials are listed in Table 7.1.

Generally, matching materials are used between the mated fibers in both mechanical and adhesive-bonded splice methods to reduce splice loss and reflection,

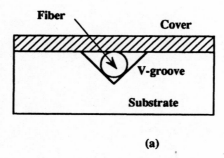

(a)

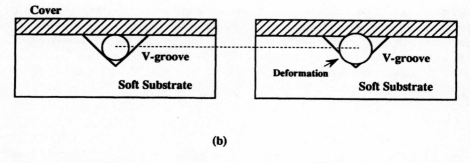

(b)

Figure 7.1 Fiber alignment method using a V-groove: (a) hard V-groove; (b) soft V-groove.

Table 7.1
Thermal Expansion Coefficient of Several Materials

Materials	Thermal Expansion Coefficient $(10^{-6}/°C)$
Silica fiber	~0.6
Silicon	2.3
Al_2O_3	~5.5
Stainless steel	~17
Plastics (silica-filled)	5–30

which, due to the Fresnel reflection, can be reduced by about 0.3 dB. Although splice loss is not sensitive to the refractive index of the matching materials, reflection is very sensitive, as discussed in Chapter 3. Therefore, accurate index matching is required to suppress the reflection. When very low reflection is required, the temperature change of the refractive index of the matching materials must be taken into consideration. Silicon gels, silicon oils, UV-curable adhesives, and epoxy resins are examples of matching materials. In the adhesive-bonded splice method, an adhesive serving both the adhesive material and the index-matching material is preferred.

A mechanical splice uses a mechanical force to sustain the fiber alignment; therefore, this splice can be respliced (reenterable). However, compared to an optical connector, the mechanical splice is generally difficult to resplice (reconnect). An adhesive-bonded splice uses an adhesive material to sustain the fiber alignment. Therefore, an adhesive-bonded splice is a permanent splice, which is similar to a fusion splice in this sense. For resplicing, a mechanical splice is similar to an optical connector and an adhesive-bonded splice is similar to a fusion splice.

7.2 MECHANICAL SPLICE

Several types of mechanical splices are explained below, and are organized according to alignment materials.

7.2.1 Silicon

A multifiber splice (array splice) was proposed and demonstrated in 1975 [1–4]. At that early investigation stage, a grooved chip used for the multifiber splice was made from metal and plastics. Later, it was made from silicon for the purpose of accurate and mass production [5,6]. An array splice for a fiber ribbon is shown in Figure 7.2 [7]. Using photolithographic and etching techniques, an accurate V-groove with a unique angle (70.5 deg in the case of (100)-oriented silicon) is formed in a silicon substrate [6]. A technique similar to that for producing ICs is used; therefore, it is suitable for mass production. The fibers in a fiber ribbon are placed between grooved silicon chips and then bonded. After bonding, fibers in silicon chips are polished to obtain good endface quality. When end preparation for two fiber ribbons is finished, the ribbons are aligned and sustained mechanically as shown in Figure 7.2.

This method is demonstrated for both multimode and single-mode fibers. For application to single-mode fibers, better dimensional control of silicon chips must be made. The stacked-fiber splice for 12 × 12 fibers has also been proposed and demonstrated [5], and it is shown in Figure 7.3. Average splice loss has been reported to range from 0.16 to 0.38 dB for multimode fibers. The stacked-fiber splice is based on the array splice using silicon chips.

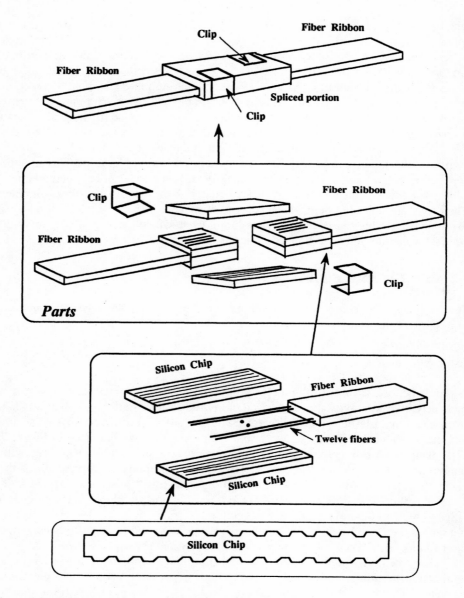

Figure 7.2 Mechanical array splice using silicon chips. (After [5–7].)

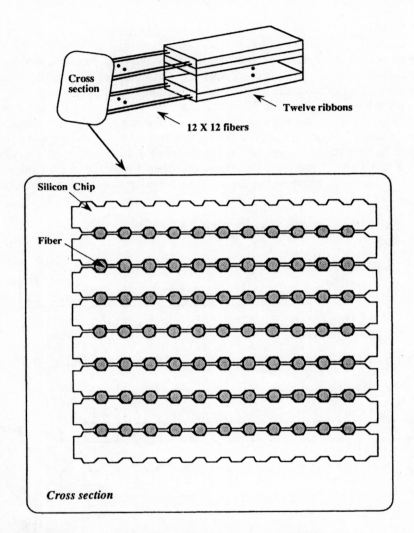

Figure 7.3 Mechanical stacked-array splice using silicon chips. (After [5].)

7.2.2 Plastic

A V-grooved substrate using plastics can be made with the precision molding technique used for mechanical splices [2,8,9]. Plastic substrates are used for mass production (low cost). Examples are shown in Figures 7.4 and 7.5. Generally, these substrates are made from glass-filled polymer resin, not from pure polymer resin. By using glass filling, the thermal expansion coefficient of a plastic substrate tends to have a lower value. In Figure 7.4, fibers are placed in grooves formed on the substrate. In Figure 7.5, fibers are placed between a cover and a plastic substrate. The rod and fastener are used to sustain the fiber alignment in two substrates in Figures 7.4, and 7.5, respectively. By using an adhesive instead of rods or a fastener, these splices can become adhesive-bonded splices.

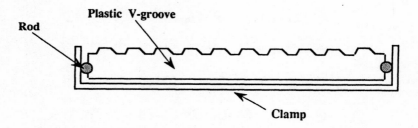

Figure 7.4 Mechanical array splice using plastic V-groove substrate. (After [8].)

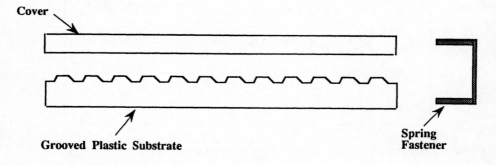

Figure 7.5 Mechanical array splice using plastic V-groove substrate. (After [9].)

7.2.3 Glass

Glass is suitable for mass production, as demonstrated by the fiber manufacturing process. Glass parts are made with drawing techniques. One example of such mechanical splices is shown in Figure 7.6. A monocoated fiber is inserted into a glass ferrule and fixed by an adhesive, and then polished, which is the same procedure used for an optical connector. In this case, a UV-curable adhesive is used to fix the fibers in the glass ferrule. With an alignment sleeve or spring fastener, two plugs are mechanically spliced [10,11]. In the case of single-mode fibers, these plugs are rotated to minimize the splice loss by using power monitoring [11]. The principle used in this mechanical splice method is similar to that used in formally proposed double eccentric connectors [12,13].

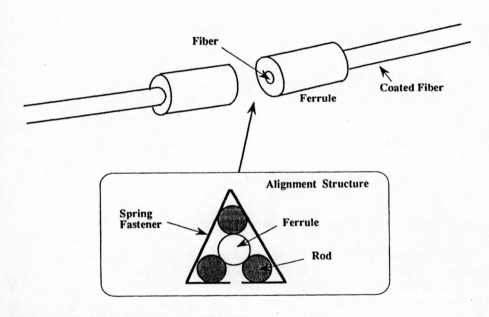

Figure 7.6 Mechanical splice. (After [10,11].)

7.2.4 Other Materials

Metal is also used for fiber alignment materials to form a V-groove in a mechanical splice. One example, designed for monocoated fibers, is in [14]. The above-mentioned mechanical splices use so-called hard materials for alignment materials. An example of using soft materials is shown in Figure 7.7 [15]. The splice method shown in this figure is designed for monocoated fibers and the elastic materials are used for a soft material. Although the structure used in Figure 7.7 is slightly different from that used in Figure 7.1(b), the principles of them are same.

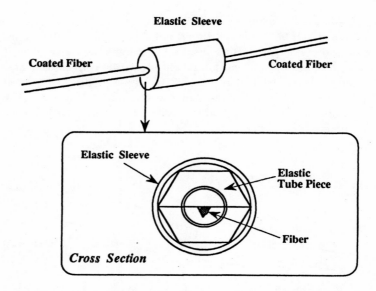

Figure 7.7 Mechanical or adhesive-bonded splice using elastic materials. (After [15].)

7.3 ADHESIVE-BONDED SPLICE

Adhesive-bonded splices are similar to mechanical splices when we change from mechanical force to adhesive force to sustain the alignment. Therefore, the mechanical splice can be the adhesive-bonded splice, if we bond with an adhesive [3]. Several materials, such as silicon, glass, plastics, and ceramics, have been investigated for adhesive-bonded splices, as for mechanical splices. Some of them are explained below according to alignment materials.

7.3.1 Plastics

A V-grooved substrate using plastics can be made with the precision molding technique used for adhesive-bonded single-fiber splices [16] and for adhesive-bonded array splices [3,17]. An example of array splices is shown in Figure 7.8, where a multi-V-grooved substrate is made from plastics. Like the substrates used for mechanical splices, these substrates are made from a glass-filled polymer resin. In these examples, epoxy resins are used for the adhesive and the index-matching materials.

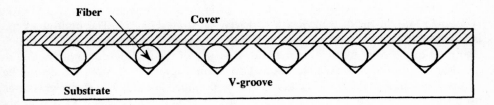

Figure 7.8 Adhesive-bonded array splice using V-grooves. (After [3,17].)

7.3.2 Glass

Adhesive-bonded splices using a glass V-groove substrate and a flat glass cover have been proposed [18]. Both the substrate and the cover are made with drawing techniques. The structure used is similar to that shown in Figure 7.1(a).

Glass is transparent so it is possible to use a UV-curable adhesive instead of a heat-curable adhesive. For making splices, a UV-curable adhesive is generally faster than a heat-curable adhesive. Bonded splices using a glass tube and a UV-curable adhesive have been proposed [19–21]. One example is shown in Figure 7.9, where LID (the one-point method) is used for the power monitoring method [19]. The monitoring method is similar to that shown in Figure 5.21(c). With the proposed power monitoring and alignment mechanisms, monocoated fibers inserted in a glass tube are aligned before bonding. The UV light source is applied at the glass tube and two mated fibers in the glass tube are adhesive-bonded. Another method also uses a glass tube and a UV-curable adhesive; the power monitoring method used in this splice method is similar to that shown in Figure 5.21(b) (the two-points method) [20,21].

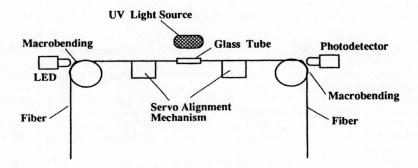

Figure 7.9 Adhesive-bonded splice using UV-curable adhesive. (After [19].)

An array splice using glass is shown in Figure 7.10, where a fused glass rod is used for alignment instead of a V-grooved substrate [22,23]. The multiglass rod is fabricated from the mother glass rods by the heating and drawing method, which is similar to the fiber fabrication method. Because of the scaling-down effect due to the drawing process, dimensional deviation or misalignment of mother rods can be reduced and an accurate multiglass rod can be obtained. In [22,23], Pyrex glass was used for the mother rod materials, and the reported splice loss was 0.11 dB on average for a five-fiber ribbon with GI fibers with a 50-µm core diameter [23]. The principle of using the rods for obtaining accurate dimensions is similar to that used in the proposed three-rod connector structure [24].

Figure 7.10 Adhesive-bonded array splice using multiglass rod. (After [22,23].)

7.3.3 Metal

A single-fiber splice method using two steel rods has been proposed [25]. Two steel rods are pressed by a fastener (an elastic element) and two fibers are placed in the groove made by the two rods. Then epoxy resin is applied to fix them. The proposed method was used for the field trial using GI fibers at an early investigation stage [25].

REFERENCES

[1] Cherin, A. H., and P. J. Rich, "A Splice Connector for Forming Linear Arrays of Optical Fibers," Topical Meeting of Optical Fiber Transmission, Williamsburg, VA, January 1975.

[2] Cherin, A. H., and P. J. Rich, "Multigroove Embossed-Plastic Splice Connector for Joining Gropes of Optical Fibers," Appl. Opt., Vol. 14, 1975, p. 3026.

[3] Chinnock, E. L., D. Gloge, P. W. Smith, and D. L. Bisbee, "Preparation of Optical Fiber Ends for Low-Loss Tape Splices," Bell Syst. Tech. J., Vol. 54, 1975, p. 471.

[4] Miller, C. M., "A Fiber-Optic-Cable Connector," Bell Syst. Tech. J., Vol. 54, 1975, p. 1547.

[5] Miller, C. M., "Fiber-Optic Array Splicing With Etched Silicon Chips," Bell Syst. Tech. J., Vol. 57, 1978, p. 75.

[6] Schroeder, C. M., "Accurate Silicon Spacer Chips for an Optical-Fiber Cable Connector," Bell Syst. Tech. J., Vol. 57, 1978, p. 91.

[7] Miller, C. M., "Mechanical Optical Splices," IEEE J. Lightwave Technol., Vol. LT-4, 1986, p. 1228.

[8] Delebecque, R., E. Chazelas, and D. Boscher, "Flat Mass Splicing Process for Cylindrical V-Grooved Cables," International Wire and Cable Symp. (IWCS), 1982, p. 178.

[9] Hardwick, N. E., III , and S. T. Davis, "Rapid Ribbon Splice for Optical Fiber Field Splicing," International Wire and Cable Symp. (IWCS), 1985, p. 255.

[10] Aberson, J. A., and K. M. Yasinski, "Multimode Mechanical Splices," European Conf. on Optical Communications (ECOC'84), 1984, p. 182.

[11] Miller, C. M., G. F. DeVeau, and M. Y. Smith, "Simple High-Performance Mechanical Splice for Single-Mode Fibers," Conf. on Optical Fiber Communication (OFC'85), MI2, 1985.

[12] Guttman, J., O. Krumpholz, and E. Pfeifer, "A Simple Connector for Glass Fiber Optical Waveguides," Vol. 29, 1975, p. 50.

[13] Tsuchiya, H., H. Nakagome, N. Shimizu, and S. Ohara, "Double Eccentric Connectors for Optical Fibers," Appl. Opt., Vol. 16, 1977, p. 1323.

[14] Patterson, R. A., "Fiber to the Home? A Lot Will Depend on How You Splice It," Telephony, 12 June 1989, p. 39.

[15] Carlsen, W. J., "An Elastic-Tube Fiber Splice," Laser Focus, April 1980, p. 58.

[16] Hirai, M., S. Seikai, and N. Kashima, "Splice of Optical Fiber and Cable," Review of ECL, Vol. 27, 1979, p. 966.

[17] Nagasawa, S., H. Murata, and T. Satake, "Study on Improvement of Splice Loss Temperature Characteristics in V-Groove Optical Fiber Mass Splice," IECE of Japan, Vol. J64-B, 1981, p. 209 (in Japanese).

[18] Tynes, A. R., and R. M. Derosier, "Low-Loss Splices for Single-Mode Fibers," Electron. Lett., Vol. 13, 1977, p. 673.

[19] Cambell, B. D., P. M. Simon, J. T. Triplett, and R. E. Tylor, "Fiber Optic Connection System," International Wire and Cable Symp. (IWCS), 1982, p. 149.

[20] Deveau, G. F., C. M. Miller, and M. Y. Smith, "Low Loss Single Mode Fiber Splices Using Ultraviolet Curable Cement," Conf. on Optical Fiber Communication (OFC'83), PD 6, 1983.

[21] Reynolds, M. P., and P. F. Gagen, "Field Splicing of Single-Mode Lightguide Cable," International Communication Conf. (ICC'84), 1984, p. 1071.

[22] Nagasawa, S., and H. Murata, "Multi-Glass-Rod Optical-Fiber Splicers Made by Drawing Technique," Electron. Lett., Vol. 16, 1980, p. 136.

[23] Nagasawa, S., and H. Murata, "Multi-Glass-Rod Optical-Fiber Splicers and Connectors Made by Drawing Technique," IECE of Japan, Vol. E66, 1983, p. 305.

[24] Metcalf, B. D., and C. W. Kleekamp, "Dual Three-Rod Connector for Single Fiber Optics," Appl. Opt., Vol. 18, 1979, p. 400.

[25] Cocito, G., B. Costa, S. Longoni, L. Michetti, L. Silverstri, D. Tibone, and F. Tosco, "COS 2 Experiment in Turin: Field Test on an Optical Cable in Ducts," IEEE Trans. Commun., Vol. COM-26, 1978, p. 1028.

Chapter 8
Optical Connectors

Optical connectors are used for connecting optical fibers without using a machine or a tool. Here, we define an optical connector as an easy reconnection component, like an electric connector. It is also possible to reconnect a mechanical splice, but it is not easy to do. A mechanical splice requires a tool for disconnecting or connecting. On the other hand, an optical connector requires no tools. A mechanical splice can be used as a connector if the housing or the clamping mechanism is redesigned for easy reconnection. Generally, an optical connector has larger connection loss compared to splices. Explained in this chapter are ordinary single-fiber connectors, array-fiber connectors, and 2D fiber connectors.

8.1 OUTLINE OF OPTICAL CONNECTOR

Many types of optical connectors have been proposed and used up to this point. Optical connectors are used inside buildings, or within closures outside. They are intended to be used with a monocoated fiber or a fiber ribbon. Optical connectors are classified in several ways. Some of their classifications are listed in Tables 8.1 and 8.2.

In Table 8.1, the classification is based on the simultaneous connection fiber number. Multifiber connectors are divided into four categories according to the unit connector, as listed in the table. One type of multifiber connector is constructed by assembling several single-fiber connectors in one housing, and a single-fiber connector is the unit connector in this case. Array-fiber connectors (array connector) are the most popular multifiber connectors, and fibers are aligned in a row (1D aligned). They are suitable for the connection of fiber ribbons. Stacked-array-fiber connectors are composed of array-fiber connectors as unit connectors. Twelve layers for a fiber ribbon with twelve fibers, resulting in a 12 × 12 mechanical splice, has been demonstrated, as explained in the previous chapter [1]. This mechanical splice can be a

---

Table 8.1
Type of Connectors

Type	Category	Dimension of Unit Connector	Characteristics	Example (Fiber Number)
Single-fiber connector	Single-fiber connector	0D	Low connection loss; suitable for monocoated fiber	1
Multifiber connector	Assembly of single-fiber connector	0D	Suitable for a monocoated fiber; low fiber density	2–12
	Array-fiber connector	1D	Suitable for a fiber ribbon	4–12
	Stacked-array-fiber connector	1D	Suitable for a fiber ribbon; high fiber density	12 × 12
	2D fiber connector	2D	Suitable for a fiber ribbon; ultrahigh fiber density	50 ~ 200

Table 8.2
Connector Classification from Mating Method Viewpoint

Type	Characteristics		Comment
	Connection Loss	Reflection	
Dry			
Air gap	○	△	Ordinary connector
Physical contact	◎	◎	Careful polishing required
Film			
Index matching film	◎	○	
Wet			
Index matching gel	◎	◎	
Index matching oil			
Lens	△	△	Large endface gap is possible

◎, Very good; ○, good; △, not very good.

stacked-array-fiber connector in principle when the alignment mechanism is rede-signed. Two-dimensionally aligned fiber connectors for both multimode and single-mode fibers were demonstrated in 1987 and 1988 [2,3]. They are very small and the fiber density in a connector plug (i.e., the fiber number per unit area of a plug) is very high. The typical fiber density is shown in Figure 8.1 for several dimensional fiber connectors as a function of the dimension. The ordinary connector for a mono-coated fiber (zero-dimensional (0D) connector) is shown as a reference. Array-fiber connectors and 2D fiber connectors have 2.5 and 10 times the fiber density, re-spectively, when compared to the 0D connector density. These high-fiber-density connectors are important for high-density packing for both optical cable-to-cable con-nection and optical cable-to-device connection when the fiber number is very high.

Several mating mechanisms have been used for these connectors. A sleeve with a slot, which serves as a fastening spring, is often used for 0D connectors. Guide pins are used for plastic mold connectors, such as array and 2D fiber connectors. The V-groove is used not only for the alignment of fibers in a fiber ribbon, but also for the mating of two plugs in the case of silicon-substrate array connectors. A clas-sification of connectors according to mating method is listed in Table 8.2.

Ordinary connectors are dry, which means there are no materials between con-nector plugs. One type of dry connector is a connector with an air gap. This type of connector has a reflection value of about -15 dB; this value is not sufficient for some applications. A connector with a physical contact (explained in Chapter 3) has excellent reflection properties. A connector using a transparent thin film between two plugs has been proposed for index matching. An example using index-matching

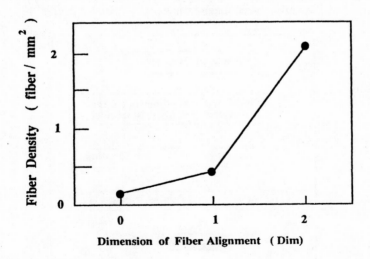

Figure 8.1 Fiber density for several connectors.

film was demonstrated in [4], and the reflection and connection loss were reported to be -25 and 0.2 dB, respectively. The reflection value of this approach is not sufficient for some applications. Index-matching gel or oil is commonly used between two plugs for reducing both connection loss and reflection value. The optical characteristics, loss and reflection, are excellent; however, some applications require no materials between plugs. A connector using a lens is similar to a dry connector with an air gap, except for the lens between plugs. By using a lens, a large endface gap is possible, which may be beneficial for some applications. However, connection loss is not low in general. Several techniques for reducing the reflection from connectors have been developed, as discussed in Chapter 3. They are summarized as follows:

1. Index-matching materials such as gels or films are inserted between plugs;
2. Physical contact is made between plugs;
3. The plug endfaces are obliquely polished.

Optical connectors are assembled using several procedures which depend on the connector structure and materials. An example of an assembly flow is shown in Figure 8.2. First, the coatings of coated fibers are removed and the fibers are cut (fiber end treatment). Then fibers are inserted into the holes of the connector ferrules or placed in V-grooves, and are then fixed by an adhesive. In some connectors, the fibers are mechanically fixed. Then the connector housing is attached. Since the fiber endface quality influences connection loss and reflection value, the fiber endfaces in a plug are generally polished. The fiber endface broken by a tool is used instead of polishing in some connectors. Polishing is done before the housing attachment in some cases. The inspection of connectors is usually done by several measuring tech-

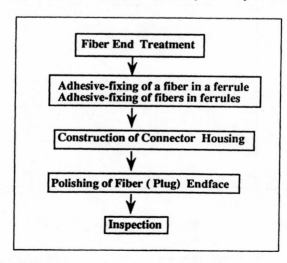

Figure 8.2 Connector assembly flow.

niques, as discussed in Chapter 4. The factors inspected are generally connection loss, reflection value, and endface quality.

8.2 SINGLE-FIBER CONNECTORS

This section discusses single-fiber connectors and multifiber connectors, which assemble single-fiber connectors as a unit connector. The shape of the plugs used for single-fiber connectors is cylindrical in most cases, but some connectors are conical [5], triangular, or a flat substrate with a V-groove. The mating mechanism for cylindrical or conical plugs uses a simple sleeve or a sleeve with a slit, as shown in Figure 8.3(a,b). It is possible to use rods or pins to align two plugs for the mating

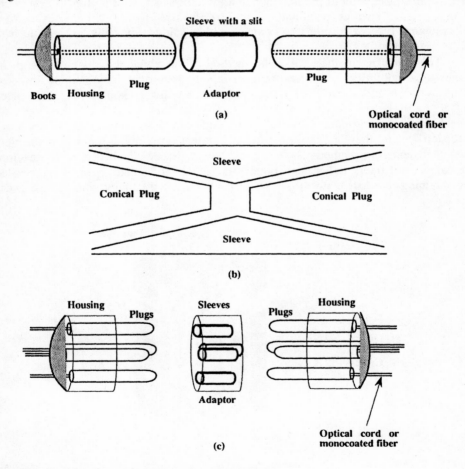

Figure 8.3 Mating mechanism in a single-fiber connector.

mechanism. One type of multifiber connectors that assembles single-fiber connectors as a unit connector is shown in Figure 8.3(c), and it uses the same mating mechanism as unit single-fiber connectors.

The plugs of single-fiber connectors are made from several types of materials, such as metal, ceramics, silicon, plastics, glass, and their compound materials. For the sake of simplicity here, we treat cylindrical or conical plugs with a hole in the center (a ferrule). A metal ferrule is formed by using a microdrill or by a pressing process. A ceramic ferrule such as an alumina ceramic or a zirconia ceramic is formed by the processes of sintering, center hole polishing, and outer surface turning [6,7]. As for a plastic plug, the plastic material flows into a mold [8–10]. One example of the manufacturing method for a single-fiber connector is shown in Figure 8.4 [10]. In this case, a mold pin is removed after the formation of a plug. A compound type of ferrule, such as metal and ceramic, is also used [6,11]. Another example of a compound type of ferrule is a connector plug using ruby balls and plastic resin [12].

The centering method (i.e., the method of matching the fiber core and the center of a cylindrical or conical ferrule) is very important for obtaining low-loss connectors. Several methods have been proposed so far, and some of them are listed in Table 8.3.

Today's typical connectors for both single-mode and multimode fibers use a precision ferrule; they belong to the nonadjustment type. Some types of connectors use other centering methods, listed in Table 8.3. Especially in the early developmental stage of fibers and optical connectors, several adjustable types were proposed for obtaining low-loss connectors. Remarkable core eccentricity existed at that de-

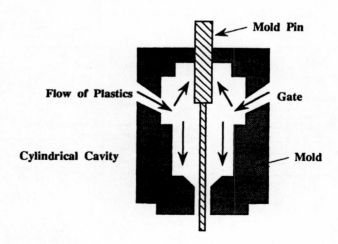

Figure 8.4 Manufacture of a plastic ferrule.

Table 8.3
Centering Method for Single-Fiber Connectors

Type	Name	Structure	Feature
No-adjustment Type	Precise Ferrule	Ferrule / Center Hole / Fiber	Precise hole is located in the center of ferrule. Fiber is inserted in this hole and fixed.
	Precise Balls or Rods	Ball or Rod / Fiber / Outer Sleeve	Balls or Rods are inserted in the outer sleeve. Fiber is inserted in the space made by these balls or rods.
	External Trimming	Sleeve for trimming / Core of Fiber	Outer surface of a sleeve is ground for trimming with referencing a fiber core as a center.
Adjustment Type	Centering of fiber core by adjustment	Outer Sleeve / Core of Fiber	Sleeve is adjusted so that a fiber core is located in the center of the sleeve. Then fixed.
	Double Eccentric Sleeve	Eccentric Sleeve / Core of Fiber	By rotaiting the inner and the outer eccentric tube, the fiber core can be located in the center of the outer tube. Then fixed.

velopmental stage. As the fiber manufacturing process improves, the nonadjustment type of connectors is commonly used, even in the case of single-mode fibers. The need for field assembly connectors has increased along with the penetration of fiber-optic transmission systems to several areas, such as LANs. Although it is possible to assemble many proposed connector types in the field, a connector with shorter assembly time is desirable. As shown in the assembly flow (Fig. 8.2), adhesive-fixing and polishing processes are required for many connector types. These procedures generally require a great deal of time. One interesting approach to this problem is a connector with a mechanical splice [13]. This structure is shown in Figure 8.5. A short length of fiber is inserted into a ferrule and fixed by an epoxy resin in the factory. The ferrule with a short fiber is prepolished in the factory. In the field, a monocoated fiber with a cleaved endface is mechanically spliced with the short fiber in the ferrule. In this approach, the epoxy curing time and polishing time are eliminated in the field. The reported connection loss is 0.3 dB with -34-dB reflection, typical for multimode fibers [13]. The drawback of this approach is the additional mechanical splice loss.

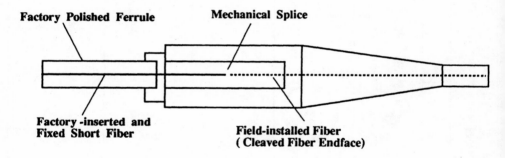

Figure 8.5 A connector with a mechanical splice. (After [13].)

8.3 ARRAY-FIBER CONNECTORS

This section discusses array-fiber connectors and stacked-array-fiber connectors, which assemble array-fiber connectors as a unit connector.

A silicon array V-groove substrate can be used for an optical connector, as described in the previous chapter. This example of silicon array V-groove connector has been developed [14,15]. The structure for one of these connectors is shown in Figure 8.6(a) [14]. It uses guide pins for the mating mechanism. The reported connection loss for a six-fiber ribbon with a 100-μm core is 0.4 dB on average when using a matching fluid [14]. The V-grooved silicon substrate and the metal block are built separately in this structure (Fig. 8.6(a)); therefore, an accurate adjustment

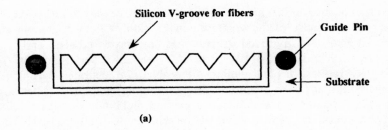

Silicon V-groove for fibers

Guide Pin

Substrate

(a)

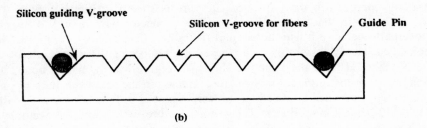

Silicon guiding V-groove

Silicon V-groove for fibers

Guide Pin

(b)

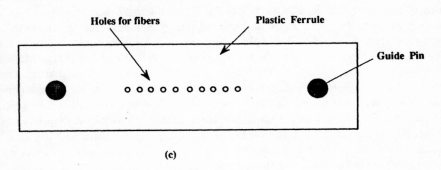

Holes for fibers

Plastic Ferrule

Guide Pin

(c)

Figure 8.6 Mating mechanism in array-fiber connectors.

is required for construction. It may be difficult to improve interchangeability between connector plugs, because of the separate construction of fiber-alignment and plug-mating mechanisms. An improved structure has been proposed [16], and it is shown in Figure 8.6(b). With the use of two types of V-grooves, fiber alignment V-grooves and guiding V-grooves for plug mating, the problem of accurate adjustment in connector construction has been overcome. Large (guiding) and small (fiber alignment) V-grooves are formed by a simultaneous etching technique on silicon chips. The reported connection loss for a six-fiber ribbon with an 80-μm core fiber is about 0.1 to 0.2 dB on average for several different connector plugs [16]. The structure shown in Figure 8.6(b) is also used for multifiber connectors using V-grooves in ceramic materials [17].

Another approach to the improved structure is shown in Figure 8.6(c). This structure is similar to that used in a single-fiber connector with regard to the connector assembly, because it uses holes, and fibers in a fiber ribbon are inserted into these holes just like the assembly of a single-fiber connector. The principle of this structure is inherited from the structure shown in Figure 8.6(b), where both fiber-alignment and plug-mating mechanisms are contained in the same substrate or plug. Array-fiber connectors using this structure were reported in [18–20] for a five-fiber ribbon with 50-μm GI core fibers. The same structure with improved accuracy in dimension is also applied for single-mode fibers in a five-fiber [21] and a ten-fiber ribbon [22]. An example of a small-sized array connector using the structure of Figure 8.6(c) is shown in Figure 8.7 [22]. To miniaturize a connector, it is first necessary to simplify its structure and secondly to miniaturize the ferrule size, considering the ferrule strength. A simple clamp spring is used to obtain a lateral force. A rectangular sleeve is simply used for protection and not for mating. Mating is realized only by two guide pins. Therefore, the rectangular sleeve is unnecessary for some applications. For fabricating these plastic ferrules, a similar molding technique shown in Figure 8.6 is used. Assembly procedures similar to those used in a single-fiber connector construction are used for these connectors because they have holes in a ferrule. Fibers in a fiber ribbon are inserted into these holes and are fixed by an adhesive. Then the plug endface is polished. Usually, these array connectors use matching materials for obtaining good connection loss and reflection properties. The reported connection loss is about 0.4 dB for a single-mode 10-fiber ribbon [22]. A connection/reconnection test was executed 1,000 times and no degradation was reported [22]. The measured crosstalk between fibers in a plug is less than −80 dB. The combination of oblique-polishing and physical-contact techniques realize a single-mode four-fiber array connector with low insertion loss (0.16 dB) and low reflection loss (−59-dB reflection) [23].

A fan-out connector for a fiber ribbon based on the structure in Figure 8.6(c) is reported in [20,24]. A fan-out connector is used for the connection between a fiber ribbon and monocoated fibers. The proposed structures are shown in Figure 8.8. The

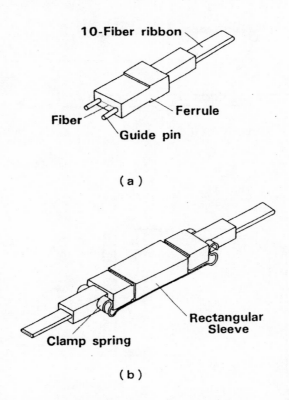

Figure 8.7 Structure of 10-fiber connector. (a) Plug; (b) mated two plugs. (After [22].)

same type of ferrules are used for both a fiber ribbon and monocoated fibers in Figure 8.8(a) [20]. Monocoated fibers are assembled and inserted into holes in a plug. The design of gathering some of the monocoated fibers is important. Another approach is shown in Figure 8.8(b) and a fan-out unit is used in this case [24]. A miniature fan-out unit is realized that considers loss and reliability due to the fiber bend [25].

Array connectors using a silicon substrate and a ferrule were reported in [26] for a single-mode fiber ribbon. The structure, a compound structure, as well as the plug construction process, is shown in Figure 8.9 (a–c). The cylindrical low-precision ferrule contains a silicon chip after assembly and its surface is ground for centering (core-centered cutting). The plug is formed to have a cylindrically convex endface. The reported connection loss is 0.2 dB without index matching [26].

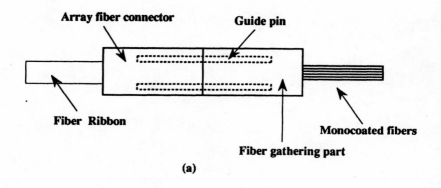

(a)

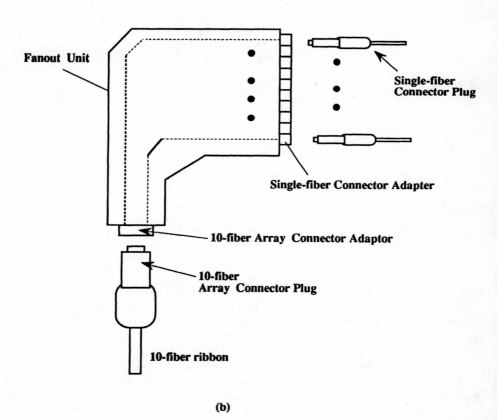

(b)

Figure 8.8 Structures of the fan-out connector: (a) direct type, (b) indirect type using a fanout unit. (After [20,24].)

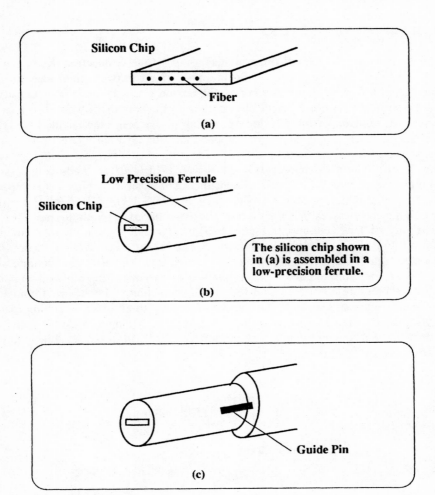

Figure 8.9 Structures of array connector: (a) silicon chip, (b) construction of a ferrule using a silicon chip, (c) construction of a plug. (After [26].)

8.4 2D-FIBER CONNECTORS

For realizing drastically small fiber connectors, 2D-fiber connectors have been proposed, where fibers in a connector ferrule are aligned in two dimensions. In order to increase the fiber number in a connector ferrule (i.e., increase fiber density), it is necessary not only for a ferrule to contain multiple fiber ribbons, but it is also necessary to miniaturize the ferrule size, taking into account the ferrule strength. In 1987, 2D 50-fiber connectors were demonstrated for GI multimode fibers with an average connection loss of 0.23 dB [2]. The structure is similar to that shown in Figure 8.6(c). The same approach is taken for 50-fiber single-mode connectors by improving the accuracy of ferrule dimensions. The basic structure of the proposed 50-fiber single-mode connector is shown in Figure 8.10(a) [3]. A plastic molded plug contains five layers of single-mode 10-fiber ribbons. The photograph of a fabricated plug endface is shown in Figure 8.10(b). To realize accurate 2D connectors, uniform filling of plastic resin into the mold space is important. When uniform filling is realized, uniform shrinkage of the resin occurs, and this results in accurate alignment. The ferrules are molded with a low-viscosity thermosetting molding compound by using a pressure-controlled transfer molding machine. The reported connection loss for fabricated single-mode 50-fiber connectors with an index-matching material is 1.2 dB on average at a wavelength of 1.3 μm [3]. A single-mode 200-fiber connector using the same approach has been reported [27]. In this case, four guide pins

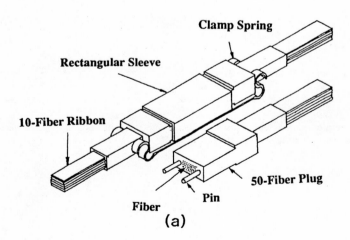

Figure 8.10 Single-mode 50-fiber connector (2D connector). (After [3].)

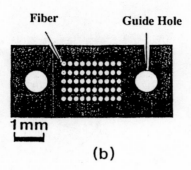

(b)

Figure 8.10 (Continued)

are used for the mating plugs. The reported connection loss is 1.4 dB on average at a 1.3-μm wavelength.

8.5 CENTERING EFFECT

8.5.1 Centering Effects in Ferrule-Type Single-Fiber Connectors

It is very important to investigate the effects of adhesive resin on connection loss in ferrule-type connectors. During assembly, the fibers are inserted into the holes of the connectors and are attached by an adhesive. Adhesive resin fills the space between the outer surface of the fiber and the hole. It is known that the location of the fiber in the hole affects connection loss. Here, we discuss the centering effect of adhesive resin on the fiber location [28]. The connector hole in a ferrule-type connector plug is shown in Figure 8.11. The center of the hole is taken as the origin of the x,y coordinate system. The symbols D, e, and $2c$ indicate the fiber diameter, the eccentricity, and the clearance, respectively. The centering effect is expected to result in a Gaussian distribution for the x,y location of the fiber axis:

$$g(x,y) = \frac{1}{2\pi\sigma^2} \exp\left(-\frac{x^2 + y^2}{2\sigma^2}\right)$$ (8.1)

Then the probability density function p of the eccentricity e is

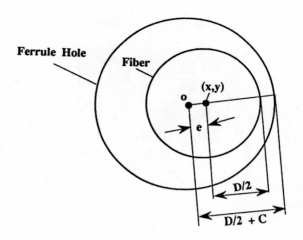

Figure 8.11 Fiber located in a ferrule hole.

$$p(e) = \int_0^{2\pi} eg(e)\, d\theta = \frac{e}{\sigma^2} \exp\left(-\frac{e^2}{2\sigma^2}\right) \qquad (8.2)$$

where e is defined as $e = (x^2 + y^2)^{1/2}$. Equation (8.2) is obtained by using $x = e \cos\theta$ and $y = e \sin\theta$. The probability distribution function P is obtained as follows:

$$P_1(e) = \int_0^e p(e)\, de = 1 - \exp\left(-\frac{e^2}{2\sigma^2}\right) \qquad (8.3)$$

Without the centering effect, a uniform distribution of the fiber location may be expected, and then the probability distribution function is as follows:

$$P_2(e) = \frac{e^2}{c^2} \qquad (8.4)$$

Experiments were conducted when the clearance $2c = 5$ μm, and the eccentricity e was measured [28]. Fifty samples of single-fiber ferrules were used in these experiments. The data are plotted in Figure 8.12 as circles. The solid lines in this figure represent theoretical values based on (8.3) and (8.4). The symbol σ in (8.3) is taken as $c/3$, because the eccentricity e ranges over $0 \le e \le c$, and $P_1(3c)$ is nearly equal to 1. These experiments show that the centering effect occurs in ferrule-type connectors under the appropriate conditions.

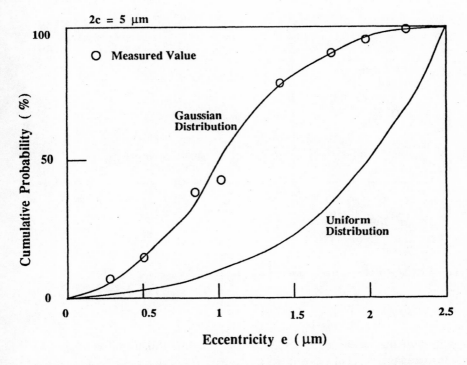

Figure 8.12 Calculated and measured cumulative provability of connection loss. (After [28].)

8.5.2 Theory of Connection Loss in Ferrule-Type Single-Mode Connectors

There are two loss factors in the case of a ferrule-type connector. One factor is caused by dimensional errors due to imperfect connector fabrication, and the other is caused by the clearance between a fiber and a ferrule hole. The major dimensional error is due to lateral displacement in most cases, so we restrict our discussion of dimensional errors to this displacement. In order to investigate the influences of dimensional error and clearance, a theory has been developed using the connector model [29].

In this theory, the outer fiber diameters vary with uniform probability distributions, and connectors have lateral displacements. The model for imperfect connectors is shown in Figure 8.13. In this model, two connectors, A and B, having the same lateral displacement d, are joined together. The origin O of the coordinates in the figure is the hole center of an ideal connector. The symbols P and Q represent the hole centers of connectors A and B. The fiber is imperfectly centered in connector

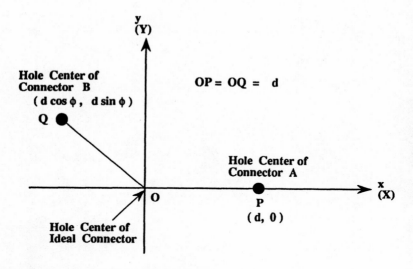

Figure 8.13 Model of nonideal connectors with lateral displacement d.

A, with a displacement having a Gaussian probability distribution of $g_1(x,y)$ around P with variance c:

$$g_1(x,y) = \frac{1}{2\pi\sigma^2} \exp\left(-\frac{(x-d)^2 + y^2}{2\sigma^2}\right) \quad (8.5)$$

The distribution of the fiber center (X,Y) in connector B is expressed as

$$g_2(X,Y) = \frac{1}{2\pi\sigma^2} \exp\left(-\frac{(X - d\cos\phi)^2 + (Y - d\sin\phi)^2}{2\sigma^2}\right) \quad (8.6)$$

where ϕ represents the azimuthal coordinate defined in Figure 8.13. The lateral displacement e among the mated fibers in connectors A and B is

$$e = ((x - X)^2 + (y - Y)^2)^{1/2} = (e_1^2 + e_2^2)^{1/2} \quad (8.7)$$

where e_1 and e_2 are defined as follows:

$$\begin{aligned} e_1 &= x - X \\ e_2 &= y - Y \end{aligned} \quad (8.8)$$

The probability distribution function $F(e)$ for lateral displacement e is

$$F(e) = \int_{c_1}^{c_2} \int_0^{2\pi} \int_0^{\theta} h(c)\, f(e_1)\, f(e_2)\, e\, de\, d\theta\, dc \tag{8.9}$$

where $e_1 = e \cos \theta$ and $e_2 = e \cos \theta$. Clearance c lies in the interval $c_1 \le c \le c_2$ with the probability distribution $h(c)$. The probability density functions $f(e_1)$ and $f(e_2)$ can be derived by using (8.5) and (8.6) and the assumption that the coordinates x and y of the fiber center are uncorrelated. The probability density function $f_L(L)$ of connection loss L can be obtained for single-mode fibers using (8.9) and the following equation [30]:

$$L = 4.34 \left(\frac{e}{\omega}\right)^2 \tag{8.10}$$

where ω is the mode field of a single-mode fiber. The results are shown in (8.11) to (8.14) [29].

$$f_L(L) = \int_{c_1}^{c_2} h(c) \frac{9\omega^2}{34.72\pi c^2} \exp\left(\frac{9d^2(\cos\phi - 1)}{2c^2}\right) \exp\left(-\frac{9\omega^2 L}{17.36c^2}\right) S\, dc \tag{8.11}$$

$$S = \int_0^{2\pi} S_1 S_2 \exp\left[-\frac{9\omega d\sqrt{L}}{2c^2\sqrt{4.34}} (\cos\theta\cos\phi + \sin\theta\sin\phi - \cos\theta)\right] d\theta \tag{8.12}$$

$$S_1 = \frac{1}{2} \mathrm{erf}[3 - 1.5(d/c)(1 - \cos\phi) + 1.5(\omega/c)\sqrt{L/4.34}\cos\theta]$$
$$+ \frac{1}{2} \mathrm{erf}[3 + 1.5(d/c)(1 - \cos\phi) - 1.5(\omega/c)\sqrt{L/4.34}\cos\theta] \tag{8.13}$$

$$S_2 = \frac{1}{2} \mathrm{erf}[1 + 1.5(d/c)\sin\phi + 1.5(\omega/c)\sqrt{L/4.34}\sin\theta]$$
$$+ \frac{1}{2} \mathrm{erf}[3 - 1.5(d/c)\sin\phi - 1.5(\omega/c)\sqrt{L/4.34}\sin\theta] \tag{8.14}$$

Using the derived equations, examples of numerical calculations are shown here [29]. The following calculations of connector losses covers two special cases. In

case I, we assume ideal fibers but admit connectors with finite dimensional tolerances. In case II, the connectors are perfect while the outer fiber diameter is allowed
to deviate from the design values within certain tolerances. In case I, the optical
fibers are ideal and the connectors are not perfect. Here, the probability density
function $f_L(L)$ is obtained using (8.11) to (8.14) with constant clearance c. The
function $f_L(L)$ is expressed simply for a small connection loss L, and this is as
follows for $\phi = 0$ and $\phi = \pi$:

$$f_L(L) = \frac{9\omega^2}{17.36c^2} \exp\left(-\frac{9\omega^2 L}{17.36c^2}\right) \text{ (for } \phi = 0) \tag{8.15}$$

$$f_L(L) = \frac{9\omega^2}{17.36c^2} \exp\left(-\frac{9\omega^2 L}{17.36c^2}\right) \exp\left(-\frac{9d^2}{c^2}\right) I_0 \left(\frac{9\omega d\sqrt{L}}{c^2\sqrt{4.34}}\right) \text{ (for } \phi = \pi) \tag{8.16}$$

where $I_0(z)$ is the modified Bessel function of z on the order of zero. In actual connectors, ϕ is possible to assume uniform distribution from 0 to 2π. The average
probability density function with respect to ϕ is defined as $f_{av}(L)$. The probability
distribution function $F_L(L)$ is obtained by the following equation:

$$F_L(L) = \int_0^L f_{av}(L)dL \tag{8.17}$$

The calculated probability distribution F for connection loss is shown in Figure 8.14.
Lateral displacement d is 0.5 and 1.0 μm, and clearance $2c$ is 1, 2, and 3 μm.
According to the result shown in Figure 8.14, the effects of clearance on connection
loss are small for large lateral displacements. However, the effects are large for small
lateral displacements.

Another example, case II, is that the connectors have perfect dimensions (ideal),
but the outer fiber diameter deviates from the design values, according to function
$h(c)$. The probability distribution function F is obtained using (8.11) to (8.14), with
lateral displacement $d = 0$ and the integration with respect to L, similar to (8.17).
Calculated results are shown in Figure 8.15. The function $h(c)$ is assumed to be a
uniform distribution from c_1 to c_2 in this calculation. Calculations are made for three
cases: $2c = 0.5$ to 2.5 μm (case A), $2c = 0.5$ to 4.5 μm (case B), and $2c = 0.5$
to 6.5 μm (case C). When fiber diameters are uniformly distributed, such as 125 $\pm$
3 μm, the probability function F of the connection loss for the ideal connector may
be similar to case C. The average connection loss value is not so different among
these three cases, as shown in Figure 8.15. However, the connection loss values for
$F = 0.8$ or $F = 0.9$ are very different for different clearance distribution.

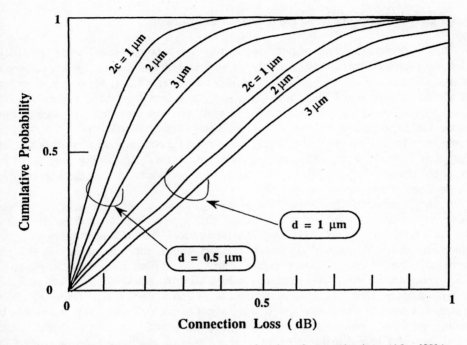

Figure 8.14 Calculated probability distribution *F* as a function of connection loss. (After [29].)

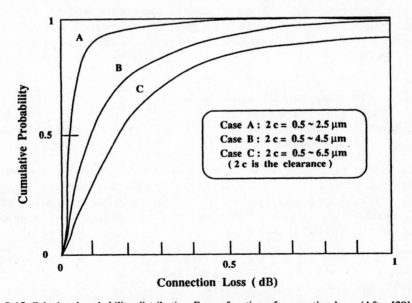

Figure 8.15 Calculated probability distribution *F* as a function of connection loss. (After [29].)

8.5.3 Centering Effect in Array-Fiber Connectors

When the centering effect occurs, connector loss becomes small even with large clearance values. The centering effect was observed by using a single-fiber connector with a clearance of $2c = 5 \ \mu$m, as described in Section 8.5.1. However, the range of clearance over which the centering effect holds is not yet been clear. Nor is it known if the centering effect exists for array-fiber connectors using a fiber ribbon. Here, the experimental investigations are shown using plastic molded array-fiber connectors [31]. The qualitative results obtained here are not general results because they may depend on the materials of an adhesive or on the assembly process. However, the tendency of the results may be helpful when considering the loss of other single-fiber or array-fiber connectors.

Single-mode 10-fiber connectors (Figure 8.7) with various clearances have been assembled. The properties of the centering effect were investigated by comparing the measured connection losses with the theoretical ones. Connectors with a displacement d $= 0.9 \ \mu$m and single-mode fibers with a core eccentricity of $0.2 \ \mu$m were used. To vary the clearance values, four kinds of fiber diameter in single-mode fiber ribbons were used. The clearance values used were $2c = 2, 3, 5,$ and $6.5 \ \mu$m. Ten-fiber connectors with these clearances were assembled. A connector with $2c = 2 \ \mu$m was used as the master plug (plug M). Connector losses were measured by mating the master plug with plugs having various clearances. The measured losses are shown in a loss distribution form and are plotted in Figure 8.16 as circles. A, B, and C represent connector plugs with $2c = 3, 5,$ and $6.5 \ \mu$m, respectively. M represents the master plug. Average connector losses for the A, B, and C plugs are 0.35 dB (sample number $N = 100$), 0.5 dB ($N = 60$), and 1.3 dB ($N = 120$), respectively. Measured values do not coincide with the calculated values using $3s = c$, except for B ($2c = 5 \ \mu$m). The assumptions used for calculations were altered as follows.

1. $5s = c$ for both plugs (M-A connection).
2. $3s = c$ for plug B and $5s = c$ for plug M (M-B connection).
3. The centering effect does not hold for plug C, and fiber positions are randomized. The effect holds only for plug M with $5s = c$ (M-C connection).

Calculations for the M-A and M-B connections are made using equations in Section 8.5.2. Calculations for the M-C connection are made using the Monte Carlo method. The calculated values are shown as solid lines in Figure 8.16, and they result in good agreement with the measured ones; therefore, the following conclusions are obtained in this case.

1. The centering effect also exists for array-fiber connectors using a fiber ribbon with normal clearances ($2c \leq 5 \ \mu$m). For smaller clearances, the centering effect tends to work more strongly.

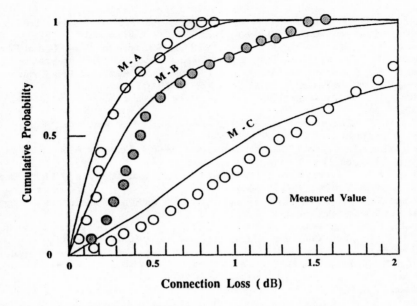

Figure 8.16 Measured and calculated probability distribution. (After [31].)

2. When clearance $2c = 6.5$ μm, the centering effect does not hold, and the fiber center locations tend to be randomly distributed.

REFERENCES

[1] Miller, C. M., "Fiber-Optic Array Splicing With Etched Silicon Chips," Bell Syst. Tech. J., Vol. 57, 1978, p. 75.

[2] Satake, T., N. Kashima, and M. Oki, "Ultra High Density 50-Fiber Connector," IEICE of Japan, Vol. E70, 1987, p. 621.

[3] Satake, T., N. Kashima, and S. Nagasawa, "Plastic Molded Single-Mode 50-Fiber Connectors," Conf. on Optical Fiber Communication (OFC'88), THJ2, 1988.

[4] Tsuchiya, H., H. Nakagome, N. Shimizu, and S. Ohara, "Double Eccentric Connectors for Optical Fibers," Appl. Opt., Vol. 16, 1977, p. 1323.

[5] Runge, P. K., and S. S. Cheng, "Demountable Single-Fiber Optic Connectors and Their Measurement on Location," Bell Syst. Tech. J., Vol. 57, 1978, p. 1771.

[6] Nawata, K., "Multimode and Single-Mode Fiber Connectors Technology," IEEE J. Quantum Electron., Vol. QE-16, 1980, p. 618.

[7] Sugita, E., R. Nagase, K. Kanayama, and T. Shintaku, "SC-Type Single-Mode Optical Fiber Connectors," IEEE J. Lightwave Technol., Vol. 7, 1989, p. 1689.

[8] Runge, P. K., "Transfer Molding of Single Optical Fiber Connectors," International Communication Conf. (ICC'79), No. 44.5, 1979.

[9] Kurokawa, T., T. Yoshizawa, S. Nara, and Y. Kitayama, "Precision Molded Fiber Connection Using an Electroformed Cavity," Electron. Lett., Vol. 17, 1981, p. 667.

[10] Satake, T., N. Kashima, and I. Sankawa, "Molding Technique for Plastic Optical Fiber Connector," IECE of Japan, Natl. Conf. Rec. on Commun., No. 2247, 1985 (in Japanese).

[11] Suzuki, N., Y. Iwahara, M. Saruwatari, and K. Nawata, "Ceramic Capillary Connector for 1. 3-μm Single-Mode Fibers," Electron. Lett., Vol. 15, 1979, p. 809.

[12] Murata, H., "Molded Optical-Fiber Connectors Using Rods and Balls," Electron. Lett., Vol. 15, 1979, p. 369.

[13] Jong, M. D., "Cleaved and Crimp Fiber Optic Connector for Field Installation," Conf. on Optical Fiber Communication (OFC'90), THA2, 1990.

[14] Fujii, Y., J. Minowa, and N. Suzuki, "Demountable Multiple Connector With Precise V-Grooved Silicon," Electron. Lett., Vol. 15, 1979, p. 424.

[15] Parzygnat, W. J., "A Multifiber Circuit Pack to Backplane Optical Connector," 6th Int. Electronics Packaging Conf. (IEPC), 1986, p. 360.

[16] Satake, T., S. Nagasawa, and H. Murata, "Low-Loss Multifiber Connectors With Plug-Guide-Grooved Silicon," Electron. Lett., Vol. 17, 1981, p. 828.

[17] Satake, T., S. Nagasawa, and H. Murata, "Multifiber Connectors Using Ceramics V-Grooves," IECE of Japan, Natl. Conf. Rec. on Commun., No. 306, 1980 (in Japanese).

[18] Tachigami, T., A. Ohtake, T. Hayashi, T. Iso, and T. Shirasawa, "Fabrication and Evaluation of a High Density Multi-Fiber Plastic Connector," International Wire and Cable Symp. (IWCS), 1983, p. 70.

[19] Satake, T., S. Nagasawa, and N. Kashima, "Design and Characteristics of Plastic Molded Multi-Fiber Connector," IECE of Japan, Vol. J68-B, 1985, p. 427 (in Japanese).

[20] Satake, T., S. Nagasawa, N. Kashima, and F. Ashiya, "Low Loss Plastic Molded Optical Multifiber Connector for Ribbon-To -Single Fiber Fan-Out," IEEE J. Lightwave Technol., Vol. LT-3, 1985, p. 1339.

[21] Satake, T., S. Nagasawa and R. Arioka, "A new type of a demountable plastic molded single mode multifiber connector," IEEE J. Lightwave Technol., Vol. LT-4, 1986, p. 1232.

[22] Satake, T., N. Kashima, and M. Oki, "Very small single-mode ten-fiber connector," IEEE J. Lightwave Technol., Vol. 6, 1988, p. 269.

[23] Nagasawa, S., Y. Yokoyama, F. Ashiya, and T. Satake, "A High-Performance Single-Mode Multi-Fiber Connector Using Oblique and Direct Endface Contact Between Multiple Fibers Arranged in a Plastic Ferrule," IEEE Photon. Technol. Lett., Vol. 3, 1991, p. 937.

[24] Nagasawa, S., T. Satake, I. Sankawa, and R. Arioka, "Optical-Fiber Fanout Connector for 10-Fiber Ribbon Cable Termination," IEEE J. Lightwave Technol., Vol. LT-4, 1986, p. 1243.

[25] Nagasawa, S., and N. Kashima, "Structural Design of Optical-Fiber Fanout Connector for Fiber-Ribbon Cable Termination," IEICE of Japan, Vol. E70, 1987, p. 276.

[26] Jacobs, A. C., G. D. Khoe, and L. J. C. Vroomen, "Single-Mode Ribbon Connector Exhibiting 0. 2 dB Loss Without Index Matching During 300 Cycles of Repeated Connection Test," European Conf. on Optical Communications (ECOC'88), 1988, p. 585.

[27] Satake, T., M. Tachikura, and Y. Negishi, "Single-Mode 200-Fiber Connector," IEICE of Japan, Natl. Conf. Rec. on Commun., No. B-696, 1989 (in Japanese).

[28] Satake, T., N. Kashima, I. Sankawa, and M. Hirai, "Effect of Resin on Fiber Centering of Ferrule-Type Optical Fiber Connectors," IECE of Japan, Vol. J67-B, 1984, p. 350 (in Japanese).

[29] Kashima, N., and T. Satake, "Relation Between Connection Loss and Single Mode Optical Fiber Diameter in a Multi-Fiber Connector," IEICE of Japan, Vol. E70, 1987, p. 1120.

[30] Marcuse, D., "Loss Analysis of Single Mode Fiber Splices," Bell Syst. Tech. J., Vol. 56, 1977, p. 703.

[31] Kashima, N., "Centering Effect in Single Mode Multi-Fiber Connectors," IEICE of Japan, Vol. E71, 1988, p. 205.

Chapter 9
Fiber Connection of Special Fibers

In this chapter, we discuss the splicing of special fibers, such as polarization-maintaining, fluoride, and EDFs. These fibers are not widely used today, but they are and will be used for special applications. Polarization-maintaining fibers are silica fibers with special structures, as explained in Chapter 1. They preserve the polarization of propagating light, and have been used for optical devices that require this preservation of polarization. In the future, they may be used in some coherent transmission systems. EDFs are silica fibers that contain erbium as a dopant. They are used for the EDFA, which is a very promising optical amplifier. When used with ordinary fibers, EDFs must be spliced or connected with ordinary fibers. Fluoride fibers have been anticipated to have ultralow loss at a wavelength around 2 μm. The silica fibers have large transmission loss over the 2-μm wavelength. Therefore, these fibers may be used when the transmission wavelength is around 2 μm. Fluoride fibers are different from silica-based fibers, such as ordinary, polarization-maintaining, and EDFs. Therefore, the melting point of fluoride fibers is very different from that of silica-based fibers. When fusion-splicing methods are used, we must seek the optimum conditions for splicing fluoride fibers.

As mentioned above, these fibers have attractive characteristics for special applications compared to the ordinary silica-based single- or multimode fibers. The splicing of these fibers is important for special applications.

9.1 CONNECTION OF POLARIZATION-MAINTAINING FIBER

Only the HE_{11} mode propagates in single-mode fibers under ordinary usage conditions. However, ordinary single-mode fibers carry two orthogonally polarized modes. The two orthogonally polarized modes are the HE_{11}^x and HE_{11}^y modes. For some applications, such as some optical measurement systems or coherent transmission systems, polarization-maintaining fibers are useful. There are many proposed polarization-maintaining fibers, as discussed in Chapter 1. Here, we mainly deal with

PANDA fibers as a typical example of these polarization-maintaining fibers. The cross section of a PANDA fiber is shown in Figure 9.1. This fiber belongs to a silica fiber and it is a polarization-maintaining single-mode fiber. The core part consists of GeO_2-doped silica glass and the cladding is pure silica glass [1]. The stress-applying part consists of SiO_2-B_2O_3-GeO_2 glass [1]. Since B_2O_3 is doped into the stress-applying part, this part has slightly different characteristics from core or cladding.

The splicing or connection of polarization-maintaining fibers is basically the same as for ordinary single-mode fibers. One exception is that precise angular alignment is required to match the principal axes of two fibers.

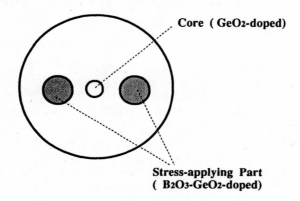

Figure 9.1 Cross section of PANDA fiber.

9.1.1 Fusion Splice of PANDA Fibers

Several angular alignments for PANDA fibers have been proposed, and they are listed in Table 9.1. The marking method uses the etched fibers for splicing. The etching rates of the undoped silica glass part and the highly doped silica glass part in a fiber are quite different, and this fact is used for the splicing of fibers [2]. For PANDA fibers, the etched stress-applying parts are used as a mark for aligning the principal axes. Using fluoric acid (HF) or the buffered HF as an etchant, a concave or convex profile is formed at a PANDA fiber endface [3]. The case for the convex profile is shown schematically in Figure 9.2(a). According to [3], a fusion-splicing loss of 0.6 dB on average was obtained when two fibers were mated in a convex-convex mating state (Fig. 9.2(b)), but an average loss of 24 dB was obtained when in a convex-concave mating state. Large splicing loss for the case of the convex-concave mating state is reported to be due to the bubble in the fusion-spliced part.

The pulse method uses the fact that the group velocity of an optical pulse depends on the principal axis position between mated PANDA fibers, due to the po-

Table 9.1
Fusion-Splicing of PANDA Fibers (Alignment of Angle θ)

Method Name	*Contents of Method*
Marking method	The stress-applying parts, which are selectively etched, are used as a mark for aligning the principal axes.
Pulse method	Short optical pulse is launched into a PANDA fiber. The monitoring of an output pulse is used for aligning the principal axes. This method uses the fact that the group velocity of pulse depends on the principal axis position between mated PANDA fibers, due to the polarization dispersion.
Extinction method	This method uses the fact that the extinction of output light from the polarization-maintaining fibers depends on the mating state. The measurement of the extinction ratio is made for each angle.
Reflection method	The reflected light from fiber end has two orthogonal components when the polarization axis of input light is inclined to the principal axis of the fiber, and one of them is detected. The angular alignment is based on the fact that the detected power depends on the angle between the fiber principal axis and the polarization axis of light.
Direct-core monitoring method	The collimated light through a fiber is detected by a camera and the light intensity distribution is obtained. Two high-intensity parts appear in the light intensity distribution in the case of PANDA fibers. The positions of two high-intensity parts depend on the fiber angle. This phenomenon is used to align the axes of PANDA fibers in this method.

larization dispersion [4]. The polarization dispersion σ_p (per unit fiber length) is expressed as

$$\sigma_p = \frac{1}{v_{gx}} - \frac{1}{v_{gy}} = \frac{n_{gx}}{c} - \frac{n_{gy}}{c} = \frac{1}{c}\left(\frac{d\beta_x}{dk} - \frac{d\beta_y}{dk}\right) \tag{9.1}$$

where $v_{gx}(v_{gy})$, $n_{gx}(n_{gy})$, and $\beta_x(\beta_y)$ are the group velocity, the group index, and the propagation constant for the x-polarization mode (y-polarization mode), respectively. The symbol k is equal to $2\pi/\lambda$ (λ is the wavelength), and c is the velocity of light. A short optical pulse is launched into a PANDA fiber. In the case of [4], a pulse with a 540-ps pulsewidth from a YAG (yttrium-aluminum-garnet) laser was used. The monitoring of an output pulse is used for aligning the principal axes, as shown in Figure 9.3. When $\theta = 0$ deg, the widest pulsewidth is obtained due to the polarization dispersion σ_p. When $\theta = 90$ deg, the shortest pulsewidth is obtained because the velocity of an optical pulse with v_{gx} is changed to the velocity with v_{gy} after splicing, and a similar phenomenon occurs for another polarized pulse. The

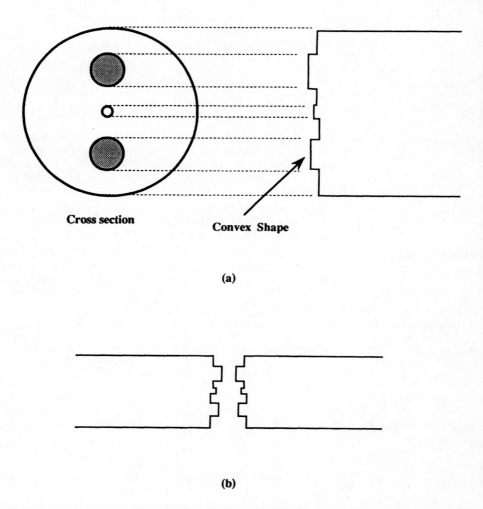

Cross section

Convex Shape

(a)

(b)

Figure 9.2 Angular alignment by marking method: (a) convex etching; (b) convex-convex mating. (After [3].)

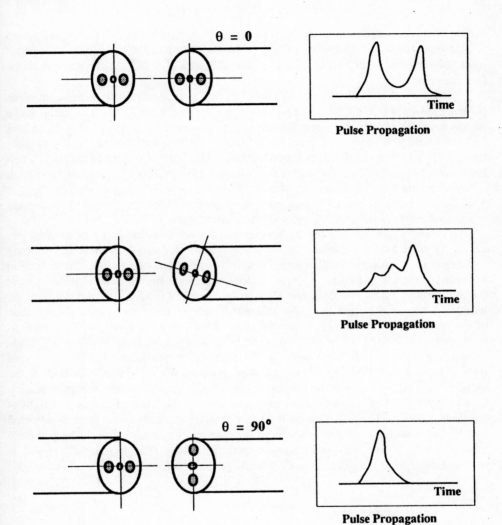

Figure 9.3 Angular alignment by pulse method. (After [4].)

averaged velocity after splicing results in the shortest pulsewidth. The obtained splice loss is reported to be about 0.5 dB [4]. This method has the drawback of using a light source with a very short optical pulse and which is difficult to use for short-fiber splicing.

The extinction method uses the fact that the extinction of output light from the polarization-maintaining fibers depends on the mating state [5,6]. The setup for a fusion splice is shown in Figure 9.4. By using this method, the reported splice loss and the extinction ratio of spliced fibers average 0.08 dB and over 30 dB, respectively [6]. The drawback of this method is that the measurement of the extinction ratio must be made for each angle, and this is time-consuming. Another drawback is based on the fact that this method is the three-points method discussed in Chapter 5. A light source, a fusion-splicing machine, and a detector must be used, and these are placed in three different locations when used in the field.

The reflection method has been proposed for overcoming some of these drawbacks [7]. This method is useful in cases where the fiber length is short and splicing and detection are located in different places. The setup and principle of the reflection method are shown in Figure 9.5(a,b). A low-coherence light source is used to treat the two orthogonal modes HE_{11}^x and HE_{11}^y as independent modes after the propagation length z_0, which is longer than the coherent length of the light source. In the case of the parameters used in [7], where an LED is used as a light source, the length z_0 is reported to be 0.1 m. A Rochon prism is used to separate the two reflected orthogonal modes; only one mode is detected by a power meter. The polarized light is input into a PANDA fiber, and some portion of the light reflects both at the input fiber end and at the output fiber end (Fig. 9.5(b)). When the polarization axis of input light is inclined to the principal axis of the fiber, the reflected light at the fiber output end has the two orthogonal components, and one of them is detected by the power meter. The detected power depends on the angle between the principal axis of the fiber and the polarization direction of light. The angular alignment is based on this principle. The following is the procedure for aligning two fiber angles [7].

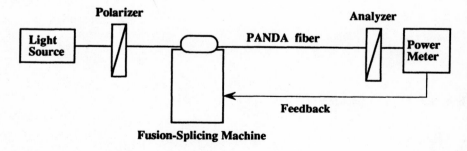

Figure 9.4 Angular alignment by extinction method. (After [6].)

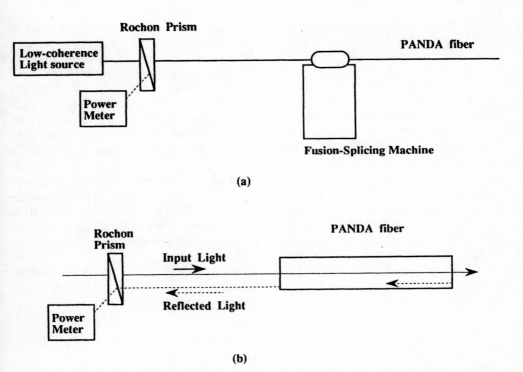

(a)

(b)

Figure 9.5 Angular alignment by reflection method. (a) Setup, (b) input and reflected lights. (After [7].)

1. Align the principal axis of incident fiber ($\theta_1 \to 0$ deg) by minimizing the reflected power. (The angle θ_1 corresponds to the angle between the principal axis of incident fiber and the input polarization direction of input light.)
2. Align $\theta_2 \to 45$ deg by maximizing the reflected power. (The angle corresponds to the angle between the principal axes of the incident fiber and the output fiber, which is mated to the incident fiber). The ordinary alignment (core alignment) is made at this stage.
3. Align two principal axes ($\theta_2 \to 0$ deg) by minimizing the reflected power.
4. Fusion splice of mated fibers.
5. Align $\theta_1 \to 45$ deg for monitoring.

This method is the two-points method as discussed in Chapter 5. Although several drawbacks of the previously mentioned methods are overcome, a drawback still exists because of the two-points method. That is, a light source, a fusion-splicing machine, and a detector must be used, and these are placed in two different locations when used in the field.

The direct-core monitoring method belongs to the visual method, as stated in Chapter 5. Core monitoring is used to align both the core and the polarization axis [8]. The collimated light through a fiber is detected by a camera and light intensity distribution is obtained. Two high-intensity parts appear in the light intensity distribution, because of the refractive-index difference between the stress-applying parts and the cladding part in the case of PANDA fibers. The positions of two high-intensity parts depend on the fiber angle. This phenomenon is used to align the axes of PANDA fibers in this method. The reported splicing loss and the extinction ratio average 0.04 dB and 33 dB, respectively, and the estimated misalignment angle averages 1.3 deg [8].

Fusion-splicing conditions for less degradation of the extinction ratio have been investigated in the fusion splicing of PANDA fibers [9]. Since the melting temperature of stress-applying parts (about 1,200°C) is different from that of ordinary silica parts (about 2,000°C) in a PANDA fiber, restriction of the melting region must be achieved. Intermittent discharge with shortened discharge durations, 0.2 seconds, has been used for this purpose [9].

For some applications, such as splicing PANDA fibers to make a depolarizer, the angular alignment of matching two principal axes is not always required. The angle of $\theta = 45$ deg is required in the case of making a depolarizer.

9.1.2 Connectors and Mechanical Splices of Polarization-Maintaining Fibers

The connectors and mechanical splices for ordinary single-mode fibers are applicable when it is possible to align the angles of two fibers. A plug with an alignment key has been proposed for the connector of a PANDA fiber, and the reported connection loss and extinction ratio average 0.6 and 33 dB, respectively [10]. The key is set by using a microscope and the accuracy is reported to be within ±1 deg [10]. Examples of mechanical splices are reported in [11,12]. The reported splice loss and extinction ratio are 1.7 dB and 33 dB, respectively, for rectangular-shaped polarization-maintaining fibers [12].

9.2 CONNECTION OF FLUORIDE GLASS FIBER

Silica fiber with a loss of 0.2 dB/km was obtained in 1979 [13], and the loss was thought to be near the minimum value. The potential of ultralow-loss fibers, loss less than 0.01 dB/km for IR optical fibers, was discussed in 1978 [14,15]. Some examples of the proposed materials are halide, chalcogenide, and heavy metal oxide. The minimum loss of fluoride glass and chalcogenide glass fibers was predicted to be 0.01 to 0.001 dB/km around the 2- to 5-μm wavelength region [16]. Using such ultralow-loss fibers, the repeaterless ultralong-distance transmission systems will be realized. Although the researchers have intensively investigated the ultralow-loss fibers, fibers with a loss less than 0.2 dB/km (loss of silica-based fibers) have not yet been realized at the time of writing (1994).

The number of reported splices and connectors for such fibers is not very large. The mechanical splices and connectors developed for silica-based fibers may be applicable for such fibers. However, the conditions of fusion splicing may be different because of the different melting temperature. The fusion splicing of fluoride fibers is reported in [17–19]. In the case of fluoride fibers, the heating temperature must be 300° to 400°C, which is much lower than the 2,000°C used for silica fibers. Also, crystallization during heating decreases the mechanical strength of fibers and increases loss [17]. To lower the heating temperature, fibers are located at an offset x from the electrode axis and short electrode gap d is used (Fig. 9.6). Two conditions, $x = 0.5$ mm for $d = 0.4$ mm and $x = 1.5$ mm for $d = 1$ mm, are used for splicing multimode fluoride fibers with a 50-μm core, 125-μm outer diameter, and $\Delta = 0.7\%$ [17]. The reported loss for these multimode fluoride fibers averages 0.14 dB. Another approach using discharge uses the condition of small discharge current (about 7 mA dc) and short heating time (0.1 to 1 second) [19]. An average splice loss of 0.08 dB and about 85% of the initial fiber strength have been obtained using an inert gas purge (helium gas) during splicing [19]. The fibers used in this experiment are multimode fluoride fibers with a 40-μm core and 125-μm outer diameter with a numerical aperture of 0.14. A fusion splice using a CO_2 laser is also possible and has been demonstrated [18]. The reported splice loss averages 0.28 dB for multimode fibers with a 54-μm core and a numerical aperture of 0.14 [18].

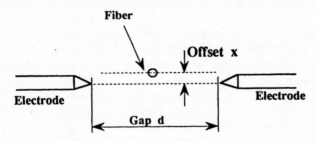

Figure 9.6 Fusion splice of fluoride fibers. (After [17].)

9.3 CONNECTION OF ERBIUM-DOPED FIBER

The EDFA will be used for many applications because of its good characteristics, such as high gain (30 to 45 dB), low noise (noise figure $F = 3$ to 4 dB), and polarization-independent gain. An EDFA has low insertion loss compared to a laser diode amplifier, because the connection loss of an EDF and an ordinary single-mode fiber is low. The typical configuration of an EDFA is shown in Figure 9.7 [20]. An EDFA can amplify light at a wavelength of around 1.5 μm, and this is one of the

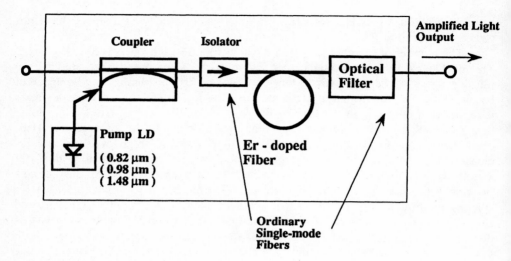

Figure 9.7 Configuration of EDFA.

wavelengths used in optical transmission systems. Amplification occurs when the pump optical power is above a certain value, which is known as the threshold pump power. Reported wavelengths for laser diode pumping are 0.82, 0.98, and 1.48 μm. An isolator is used in Figure 9.7 for stable amplification by suppressing the feedback from a signal light and an amplified spontaneous emission (ASE). An EDF is basically silica fibers with erbium as a dopant. Since an EDFA belongs to a three-level system, absorption occurs for the weak pump power case. Therefore, the erbium dopant concentration in the center part of the core is an important technique for obtaining the high gain coefficients for pump power. As a result, an EDF has a different MFD from ordinary single-mode fibers or DSFs. For example, MFDs for DSFs and EDFs are 8 and 4 μm, respectively, at a wavelength of 1.55 μm. When used with ordinary fibers, EDFs must be spliced or connected with ordinary fibers. Splice loss due to the mismatch of MFDs is calculated to be about 1.9 dB by using the equation discussed in Chapter 3. This value is very large compared to the ordinary splice loss of single-mode fibers. Several local MFD transforming methods can be used to lower splice or connection loss. These methods were initially proposed for other applications. According to [21], they are classified as follows:

1. Tapering a single-mode fiber by stretching the fusion-spliced portion (down-tapering) [22];
2. Tapering a preform by drawing (up-tapering) [23];
3. Diffusion of dopant in a single-mode fiber by heat treatment [24].

The methods of fabricating a taper are shown in Figure 9.8, and these local MFD transforming methods are explained below.

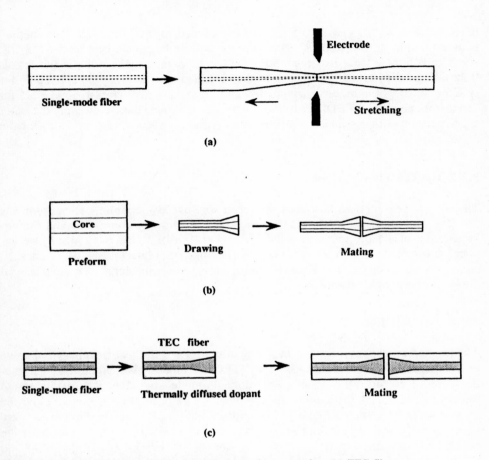

Figure 9.8 Fabrication of tapers: (a) down-tapering; (b) up-tapering; (c) TEC fiber.

9.3.1 Taper Splicing Method (Down-Tapering)

The taper splicing method was proposed originally for decreasing fusion-splicing loss for single-mode fibers in 1978 [22]. After the usual fusion splice, the spliced part is heated again by an electric discharge and it is stretched by pulling (Fig. 9.8(a)). The softened spliced part develops a tapered shape by heating and pulling, and both the core diameter 2a and the core misalignment x reduce to $R(2a)$ and Rx, respectively, where R is the ratio of the core diameter at the spliced point to the original core diameter 2a. Two major effects, such as the decreasing of core misalignment and the spreading of MFD due to smaller core diameter, result in lower splice loss when compared to the original nontapered splice. The improvement of reducing angular offset is also pointed out in [22]. It has also been pointed out that, with this taper splice, a splice loss of 0.5 dB can be reduced to 0.05 dB [22]. This method is used for splicing single-mode fibers with a relatively large core eccentricity [25,26]. Average fusion-splicing loss reductions have been reported where the initial loss is 0.38 to 0.24 dB [25] and 0.40 to 0.25 dB [26]. This method may be applicable for splicing an EDF and an ordinary single-mode fiber, and may be also applicable for connectors between an EDF and an ordinary single-mode fiber. However, the outer diameter is also taper-shaped, and this may require a special plug (ferrule) for the connector.

9.3.2 Up-Tapering Method

The up-tapering method has been proposed for low-loss coupling of components [23,27]. The up-taper is fabricated at the stage of drawing a preform, and the enlarged core of a taper transforms the MFD (Fig. 9.8(b)). This method may be applicable for mechanical splicing, bonded splicing, or connectors between an EDF and an ordinary single-mode fiber. However, it requires a special preform and special plugs (ferrules) or hardware.

9.3.3 TEC Method

The method of dopant diffusion in a single-mode fiber by heat treatment is known as the thermally diffused expanded core (TEC) method. This method uses the phenomenon of dopant diffusion in a heated single-mode fiber (Fig. 9.8 (c)). Through the process of dopant diffusion, the core diameter 2a becomes large locally and the relative refractive-index difference Δ becomes small locally compared to the ordinary fiber part. By considering that the normalized frequency V is nearly constant, the MFD 2ω is proportional to the core diameter 2a. This is known from (2.28) and (2.29). That is, 2ω is expanded by the taper formed by dopant diffusion. An example of a GeO_2-doped core fiber (silica fiber) is demonstrated in [24,21]. An example of a fiber with a pure silica core and fluorine-doped cladding is discussed in [28,29]. TEC fiber is fabricated by a furnace [21] or by a flame from a microburner [30]. Fusion splicing single-mode fibers with two different MFDs, 4.5 and 7.8 μm, has been demonstrated by using TEC fiber technology. The reported fusion-splicing loss reduction to about 0.1 dB, from the initial 1-dB splice loss, resulted from heating the fiber with microburners for several minutes [30]. The TEC method can be used to splice an EDF and an ordinary single-mode fiber. In the case of a discharge fusion splice, heating with only an electric discharge is required, not a furnace or a microburner. In this case, fusion splices with a longer discharge time and with additional discharges several times after fusion splicing are used [31]. It is reported that the TEC method is applicable for reducing several coupling losses, such as those in fiber-to-laser diode coupling [32] and fiber-to-fiber connections [33].

An approach other than down-tapering, up-tapering, and TEC methods has been proposed [34]. In this proposed method, one or more intermediate fibers are used to reduce the mismatch of the MFD between an EDF and an ordinary single-mode fiber.

REFERENCES

[1] Shibata, N., Y. Sasaki, K. Okamoto, and T. Hosaka, "Fabrication of Polarization-Maintaining and Absorption-Reducing Fibers," IEEE J. Lightwave Technol., Vol. LT-1, 1983, p. 38.

[2] Svaasand, L. O., S. Hopland, and A. P. Grande, "Splicing of Optical Fibers With a Selective Etching Technique," European Conf. on Optical Communications (ECOC'78), 1978, p. 304.

[3] Noda, J., N. Shibata, T. Edahiro, and Y. Sasaki, "Splicing of Single Polarization-Maintaining Fibers," IEEE J. Lightwave Technol., Vol. LT-1, 1983, p. 61.

[4] Sasaki, Y., N. Shibata, and J. Noda, "Splicing of Single-Polarization Fibers by an Optical Short-Pulse Method," Electron. Lett., Vol. 18, 1982, p. 997.

[5] Tokiwa, H., K. Mochizuki, and H. Wakabayashi, "Joint Characteristics Between Polarization-Maintaining Single-Mode Fibers," Electron. Lett., Vol. 19, 1983, p. 485.

[6] Suzuki, M., Y. Kikuchi, T. Yamada, and O. Watanabe, "Arc Fusion Splicing Machine for Single Polarization Single Mode Optical Fibers," European Conf. on Optical Communications (ECOC'83), 1983, p. 117.

[7] Kato, Y., "Fusion Splicing of Polarization Preserving Fibers," Appl. Opt., Vol. 24, 1985, p. 2346.

[8] Taya, H., K. Ito, T. Yamada, and M. Yoshinuma, "New Splicing Method for Polarization-Maintaining Fibers," Conf. on Optical Fiber Communication (OFC'89), THJ2, 1989.

[9] Ishikura, A., Y. Kato, T. Abe, and M. Miyauchi, "Optimum Fusion Splice Method for Polarization-Preserving Fibers," Appl. Opt., Vol. 25, 1986, p. 3455.

[10] Tamaki, Y., K. Koyama, H. Furukawa, O. Watanabe, and H. Yokosuka, "A Study of Connector for Polarization-Maintaining Optical Fiber," IECE of Japan, Natl. Conf. Rec. on Commun., No. 2180, 1986 (in Japanese).

[11] Sears, F. M., C. M. Miller, W. A. Vicory, and D. N. Ridgway, "Enhanced Rotary Mechanical Splice for Rectangular Polarization-Maintaining Fibers," Conf. on Optical Fiber Communication (OFC'89), THJ1, 1989.

[12] Sears, F. M., "A Passive Mechanical Splice for Polarization-Maintaining Fibers," IEEE J. Lightwave Technol., Vol. 7, 1989, p. 1494.

[13] Miya, T., Y. Terunuma, T. Hosaka, and T. Miyashita, "Ultimate Low-Loss Single-Mode Fiber at 1. 55 μm," Electron. Lett., Vol. 15, 1979, p. 106.

[14] Pinnow, D. A., A. L. Gentile, A. G. Standlee, and A. Timper, "Polycrystalline Fiber Optical Waveguide for Infrared Transmission," Appl. Phys. Lett., Vol. 33, 1978, p. 28.

[15] Van Uitert, L. G., and S. H. Wemple, "$ZnCl_2$ Glass: A Potential Ultra Low-Loss Optical Fiber Material," Appl. Phys. Lett., Vol. 33, 1978, p. 57.

[16] Shibata, S., M. Horiguchi, K. Jinguji, S. Mitachi, K. Kanamori, and T. Manabe, "Prediction of Loss Minima in Infra-red Optical Fibers," Electron. Lett., Vol. 17, 1981, p. 775.

[17] Tachikura, M., T. Satake, and Y. Ohishi, "Discharge Fusion Splicing of Fluoride Optical Fibers," IEICE of Japan, Vol. E71, 1988, p. 736.

[18] Rivoallan, L., and J. Y. Guilloux, "Fusion Splicing of Fluoride Glass Optical Fiber With CO_2 Laser," Electron. Lett., Vol. 24, 1988, p. 756.

[19] Harbison, B. B., W. I. Roberts, and I. D. Aggarwal, "Fusion Splicing of Heavy Metal Fluoride Glass Optical Fibers," Electron. Lett., Vol. 25, 1989, p. 1214.

[20] Kashima, N., "Optical Transmission for the Subscriber Loop," Norwood, MA: Artech House, 1993.

[21] Shiraishi, K., Y. Aizawa, and S. Kawakami, "Beam Expanding Fiber Using Thermal Diffusion of the Dopant," IEEE J. Lightwave Technol., Vol. 8, 1990, p. 1151.

[22] Furuya, K., T. C. Chong, Y. Suematsu, "Low-Loss Splicing of Single-Mode Fibers by Tapered-Butt-Joint Method," IECE of Japan, Vol. E61, 1978, p. 957.

[23] Amitay, N., H. M. Presby, F. V. Diamarcello, and K. T. Nelson, "Single-Mode Optical Fiber Tapers for Self-Aligned Beam Expansion," Electron. Lett., Vol. 22, 1986, p. 702.

[24] Shigihara, K., K. Shiraishi, and S. Kawakami, "Modal Field Transforming Fiber Between Dissimilar Waveguides," J. Appl. Phys., Vol. 60, 1986, p. 4293.

[25] Taya, H., T. Tanaka, T. Yamada, and O. Watanabe, "Low Splice Loss of Single Mode Fiber for Tapered Joint Method," IECE of Japan, Technical Report OQE85-110, 1985 (in Japanese).

[26] Ishikura, A., Y. Kato, and M. Miyauchi, "Taper Splice Method for Single-Mode Fibers," Appl. Opt., Vol. 25, 1986, p. 3460.

[27] Amitay, N., H. M. Presby, F. V. Diamarcello, and K. T. Nelson, "Optical Fiber Tapers—A Novel Approach to Self-Aligned Beam Expansion and Single-Mode Hardware," IEEE J. Lightwave Technol., Vol. LT-5, 1987, p. 70.

[28] Botham, C. P., "Theory of Tapering Single-Mode Optical Fibers by Controlled Core Diffusion," Electron. Lett., Vol. 24, 1988, p. 243.

[29] Harper, J. S., C. P. Botham, and S. Hornung, "Tapers in Single-Mode Optical Fiber by Controlled Core Diffusion," Electron. Lett., Vol. 24 1988, p. 245.

[30] Hanafusa, H., M. Horiguchi, and J. Noda, "Thermally-Diffused Expanded Core Fibers for Low-Loss and Inexpensive Photonic Components," Electron. Lett., Vol. 27, 1991, p. 1968.

[31] Tam, H. Y., "Simple Fusion Splicing Technique for Reducing Splicing Loss Between Standard Singlemode Fibers and Erbium-Doped Fiber," Electron. Lett., Vol. 27 1991, p. 1597.

[32] Kato, K., I. Nishi, K. Yoshino, and H. Hanafusa, "Optical Coupling Characteristics of Laser Diode to Thermally Diffused Expanded Core Fiber Coupling Using an Aspheric Lens," IEEE Photon. Technol. Lett., 1991, p. 469.

[33] Haibara, T., T. Nakashima, M. Matsumoto, and H. Hanafusa, "Connection Loss Reduction by Thermally-Diffused Expanded Core Fiber," IEEE Photon. Technol. Lett., 1991, p. 348.

[34] Holmes, M. J., F. P. Payne, and D. M. Spirit, "Matching Fibers for Low Loss Coupling Into Fiber Amplifiers," Electron. Lett., Vol. 26 1990, p. 2102.

PART III
COMPONENT WITH FUNCTION

Chapter 10
Optical Coupler and Branch

Gathering and distributing optical lightwaves are important functions for several applications. Directional and star couplers, which have these functions, are treated in this chapter. Ideal directional and star couplers are wavelength-insensitive devices. (Wavelength-selective devices such as WDM couplers are discussed in Chapter 11.) Directional couplers are often used in bidirectional transmission systems, and they are also used for mixing or branching optical signals (waves) in several transmission and measurement systems. Star couplers are used for many optical signals (waves) to mix or branch. Their applications are LAN and video distribution systems.

10.1 DIRECTIONAL COUPLER

10.1.1 Scattering Matrix and Classification

Several types of directional couplers have been used and they are classified into three major types: bulk, fused-fiber, and waveguide (WG). The bulk-type coupler uses a beam splitter, such as a half mirror ((Fig. 10.1(a)), and the other two types use a coupled-waveguide structure (Fig. 10.1(b)). The fused-fiber coupler uses fused fibers as a coupled waveguide. In the case of the waveguide-type coupler, coupled waveguides are formed in a planar waveguide.

A model of the directional coupler, which is universally used and independent of the type, is shown in Figure 10.1(c). The numbers shown in the figure are the port numbers. Using this model, the directional coupler can be expressed with a scattering matrix (discussed in Chapter 1)

$$\mathbf{b} = [S]\mathbf{a} \tag{10.1}$$

where $\mathbf{a} = (a_1, a_2, \ldots, a_i, a_n)$ and $\mathbf{b} = (b_1, b_2, \ldots, b_i, b_n)$ are the vectors, and

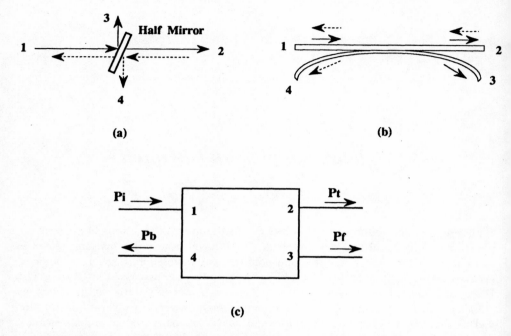

Figure 10.1 Directional couplers and model: (a) bulk-type coupler; (b) fused-fiber coupler/WG-type coupler; (c) model.

the notations a_i and b_i represent the complex amplitudes for the input and the reflected lightwaves, respectively. The notation $[S]$ is the scattering matrix. In the case of directional couplers, there are four port components and $[S]$ is expressed as

$$[S] = \begin{bmatrix} s_{11} & s_{12} & s_{13} & s_{14} \\ s_{21} & s_{22} & s_{23} & s_{24} \\ s_{31} & s_{32} & s_{33} & s_{34} \\ s_{41} & s_{42} & s_{43} & s_{44} \end{bmatrix} \tag{10.2}$$

Coupling factor C, directivity D, and isolation I are defined by using the powers shown in Figure 10.1(c):

$$C = 10 \log \frac{P_i}{P_f} \tag{10.3}$$

$$D = 10 \log \frac{P_f}{P_b} \tag{10.4}$$

$$I = 10 \log \frac{P_i}{P_b} \tag{10.5}$$

The powers are expressed with the scattering matrix; therefore, coupling factor C, directivity D, and isolation I are expressed with the scattering matrix, such that

$$C = 10 \log \frac{1 - |S_{11}|^2}{|S_{31}|^2} \tag{10.6}$$

$$D = 10 \log \frac{|S_{31}|^2}{|S_{41}|^2} \tag{10.7}$$

$$I = 10 \log \frac{1 - |S_{11}|^2}{|S_{41}|^2} \tag{10.8}$$

With the ideal directional coupler, the insertion loss is zero and there are no reflections at port 1,2. When the couplers are lossless, the scattering matrix is known to be a unitary matrix. Using the unitary matrix and reciprocity, the scattering matrix for the ideal directional coupler is

$$[S]_{\text{ideal}} = \begin{bmatrix} 0 & c_1 & jc_2 & 0 \\ c_1 & 0 & 0 & jc_2 \\ jc_2 & 0 & 0 & c_1 \\ 0 & jc_2 & c_1 & 0 \end{bmatrix} \tag{10.9}$$

where c_1, and c_2 are the real constants and the following equation holds.

$$c_1^2 + c_2^2 = 1 \tag{10.10}$$

In this ideal coupler, both D and I tend to infinity. For an ideal 3-dB directional coupler, the constants are

$$c_1 = c_2 = \frac{1}{\sqrt{2}} \tag{10.11}$$

10.1.2 Coupled-Wave Equation

The principle of directional couplers for both fused-fiber and waveguide types is the coupling of light. The directional couplers are analyzed by the coupled-wave equation. The cross section of two parallel waveguides WG_1 and WG_2 is modeled in Figure 10.2. The field of these waveguides can be analyzed using the perturbation method when the coupling of two waveguides is weak. That is, the fields E and H are expressed by the superposition of the unperturbed field of each waveguide.

$$E = A_1(z)E_1 + A_2(z)E_2 \qquad (10.12)$$

$$H = A_1(z)H_1 + A_2(z)H_2 \qquad (10.13)$$

where E_1 and H_1 are the fields for WG_1, and E_2 and H_2 are the fields for WG_2. The fields E and H satisfy the Maxwell equation, and the following equations are approximately obtained by neglecting small terms [1].

$$\frac{dA_1(z)}{dz} = j\kappa_1 A_2(z)\, e^{j(\beta_1 - \beta_2)z} \qquad (10.14)$$

$$\frac{dA_2(z)}{dz} = j\kappa_2 A_1(z)\, e^{-j(\beta_1 - \beta_2)z} \qquad (10.15)$$

where β_1 and β_2 are the phase constants (time dependence of $\exp[j(\omega t - \beta z)]$ is assumed) for WG_1 and WG_2, respectively. These equations express the coupling situation between modes with β_1 and β_2 in two waveguides. The constants κ_1 and κ_2

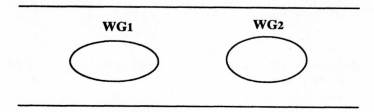

Figure 10.2 Cross section in directional couplers.

are the coupling coefficients. Here, we introduce the complex amplitudes $a_1(z)$ and $a_2(z)$:

$$a_1(z) = A_1(z)e^{-j\beta_1 z} \tag{10.16}$$

$$a_2(z) = A_2(z)e^{-j\beta_2 z} \tag{10.17}$$

Then the well-known coupled-wave equations for two waveguides are derived as

$$\frac{da_1}{dz} = -j\beta_1 a_1 + j\kappa_1 a_2 \tag{10.18}$$

$$\frac{da_2}{dz} = j\kappa_2 a_1 - j\beta_2 a_2 \tag{10.19}$$

For lossless waveguides, the equation $\kappa_1 = \kappa_2 \ (=\kappa)$ holds in the case of real κ values. This is derived by the following power conservation equation for lossless waveguides [1]:

$$\frac{d}{dz}[|a_1|^2 + |a_2|^2] = 0 \tag{10.20}$$

This equation is rewritten using (10.18) and (10.19) as

$$\frac{d}{dz}[|a_1|^2 + |a_2|^2] = \frac{d}{dz}[a_1 a_1^* + a_2 a_2^*]$$

$$= 2\text{Re}[j(\kappa_2 - \kappa_1^*)a_1 a_2^*] \tag{10.21}$$

where z^* and $\text{Re}[z]$ are the complex conjugate of z and the real part of z, respectively. Equation (10.21) must hold for any values of a_1 and a_2; then $\kappa_1 = \kappa_2 \ (=\kappa)$ holds in the case of real κ values. The solutions of the above coupled-wave equations for $a_1(0) = A$ and $a_2(0) = 0$ are

$$P_1(z) = |a_1(z)|^2 = A^2[1 - (\kappa/\Delta\beta)^2 \sin^2(\Delta\beta z)] \tag{10.22}$$

$$P_2(z) = |a_2(z)|^2 = A^2(\kappa/\Delta\beta)^2 \sin^2(\Delta\beta z) \tag{10.23}$$

$$\Delta\beta \equiv \left[\kappa^2 + \left(\frac{\beta_1 - \beta_2}{2} \right)^2 \right]^{1/2} \qquad (10.24)$$

Since the waveguides are lossless, the following equation holds from (10.22) and (10.23).

$$P_1(z) + P_2(z) = A^2 \text{ (constant)} \qquad (10.25)$$

For the coupling of the same mode in identical waveguides, $\beta_1 = \beta_2 \ (=\beta)$ holds. The solutions are

$$a_1(z) = A \cos(\Delta\beta z) \exp(-j\beta z) \qquad (10.26)$$

$$a_2(z) = jA \sin(\Delta\beta z) \exp(-j\beta z) \qquad (10.27)$$

$$\Delta\beta = |\kappa| \qquad (10.28)$$

$$P_1(z) = |a_1(z)|^2 = A^2 \cos^2(\Delta\beta z) \qquad (10.29)$$

$$P_2(z) = |a_2(z)|^2 = A^2 \sin^2(\Delta\beta z) \qquad (10.30)$$

When $\Delta\beta z = \pi/2$ (i.e., $|\kappa|z = \pi/2$), the initial power of WG_1 is completely transferred to WG_2. The physical image of the power transfer from WG_1 to WG_2 is shown in Figure 10.3 [2]. The symmetric and antisymmetric modes are used for the explanation. The symmetric mode has the same phases of amplitudes for the two waveguides WG_1 and WG_2, and the antisymmetric mode has opposite phases, as indicated in the Figure 10.3 ($z = 0$). These symmetric and antisymmetric modes are assumed to be equally excited at $z = 0$, and the field of WG_1 only exists as the result of the summation of the two modes. After traveling the length satisfying $\Delta\beta z = \pi/2$ in a coupler, the sign of the antisymmetric mode is reversed and the sign of the symmetric mode is unchanged. As a result of the summation of the two modes, the field of WG_2 only exists at the length z that satisfies $|\kappa|z = \pi/2$. $P_1(\pi/4\kappa) = P_2(\pi/4\kappa) = 0.5P_1(0)$ can also be derived, which is the case for a 3-dB coupler. The coupling length for a 3-dB coupler is used for both fused-fiber and waveguide types. The coupling coefficient κ depends on the waveguide structure (e.g., the distance of two cores, the shape of a core, and refractive-index profiles). By designing the structural parameters appropriately, a desirable κ value can be obtained. Nearly the same structures, except for the different design parameters in a fused-fiber or waveguide coupler, are used for a WDM device, which will be discussed in Chapter 11.

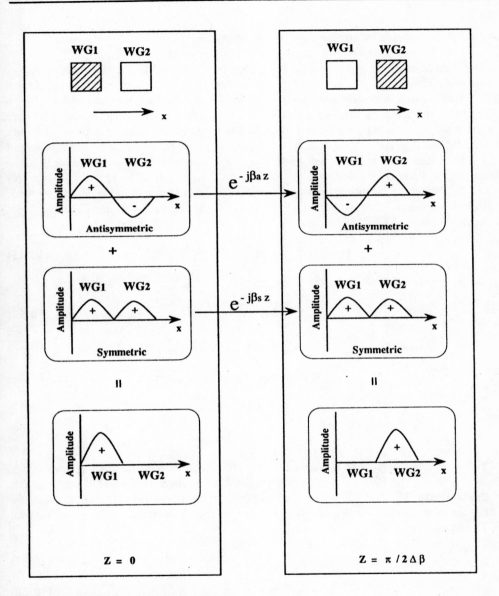

Figure 10.3 Principle of directional couplers.

10.1.3 Directional Couplers With a Rectangular Core

Directional couplers with a rectangular core in a waveguide are modeled in Figure 10.4(a). This figure is a cross section (x, y plane) having two rectangular cores with a refractive index n_1 and a cladding with n_2. Each core has the dimensions of $a \times a$. The separation of the two cores is c. These types of directional couplers were analyzed by Marcatili in the early stage of investigation on optical circuits [2]. They can be considered as a model for waveguide-type directional couplers, which will be discussed in more detail later on. The coupling coefficient of the coupler shown in Figure 10.4(a) is calculated by the theory proposed in [2]. This theory ignores the fields and the boundary conditions of the dashed regions shown in Figure 10.4(b). For example, the coupling coefficient κ for the fundamental mode E_{11} with y polarization is

$$|\kappa| = 2Ak_x^2[1 - (k_xA/\pi)^2]^{1/2} \exp\{-\pi c[1 - (k_xA/\pi)^2]^{1/2}/A\}/(\pi a k_z) \quad (10.31)$$

where A, k_x, and k_z are expressed as

$$A = \lambda/2 \, (n_1^2 - n_2^2)^{1/2} \tag{10.32}$$

$$k_x = \pi/(a + 2A/\pi) \tag{10.33}$$

$$k_z = (k_1^2 - k_x^2 - k_y^2)^{1/2} \tag{10.34}$$

$$k_1 = (2\pi/\lambda)n_1 \tag{10.35}$$

$$k_y = \pi/(a + 2An_2^2/\pi n_1^2) \tag{10.36}$$

The coupling coefficient κ shown in (10.31) is obtained from the physical image shown in Figure 10.3. The coupling length L is defined as a z value that satisfies the equation $\Delta\beta z = \pi/2$. After traveling the coupling length L, the sign of the antisymmetric mode is reversed and the sign of the symmetric mode is unchanged. Therefore,

$$\beta_a L = \pi \tag{10.37}$$

$$\beta_s L = 2\pi \tag{10.38}$$

where β_s and β_a are the propagation constants for the symmetric and antisymmetric modes. The coupling coefficient κ is expressed using (10.37), (10.38), and $|\kappa|L = \pi/2$. That is,

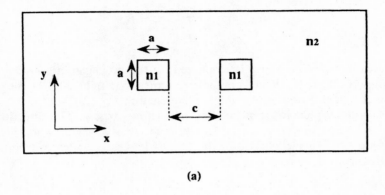

(a)

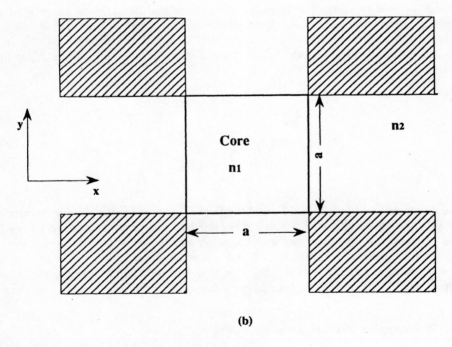

(b)

Figure 10.4 Directional coupler with a rectangular core.

$$|\kappa| = \frac{\pi}{2L} = \frac{\beta_s - \beta_a}{2} \tag{10.39}$$

This equation implies that κ is expressed by the propagation constant difference. Equation (10.31) is obtained from the electromagnetic field calculations of the propagation constants based on the above-mentioned theory [2].

Another approach for calculating κ is shown below. The directional coupler shown in Figure 10.4(a) is a case in which the two waveguides are identical. The coupling coefficients κ_1 and κ_2 for the case of lossless identical waveguides are expressed as [1]

$$\kappa_1 = \kappa_1 = \kappa = -\frac{\omega\varepsilon_0}{4P} \int_{-\infty}^{\infty} \int_{-\infty}^{\infty} [n(x, y)^2 - n_2(x, y)^2] E_2^* E_1 \, dx \, dy \tag{10.40}$$

where P is the power of the mode in each waveguide. The refractive indexes $n(x, y)$ and $n_2(x, y)$ represent the refractive-index profiles for the total waveguide (the coupler) and that for waveguide 2, respectively. This equation for the coupling coefficient is obtained through the process derived from (10.12) through (10.15). In the case of the fundamental mode E_{11} with y polarization ($H_y = 0$, where H_y is the y component of magnetic field H), κ is

$$|\kappa| = \frac{\omega\varepsilon_0(n_1^2 - n_2^2) \int_{-a/2}^{a/2} \int_{-a/2}^{a/2} E_{2y}^* E_{1y} \, dx \, dy}{2 \int_{-\infty}^{\infty} \int_{-\infty}^{\infty} E_{2y}^* H_{2x} \, dx \, dy} \tag{10.41}$$

The dominant fields are E_y and H_x and the field E_x is neglected in this approximation. By substituting these fields, which are obtained by the above-mentioned theory [2], into (10.41), the coupling coefficient κ is obtained. The value of L from the calculated κ is around 2 mm for a 7.5×7.5 μm rectangular core, $c = 3$ to 4 μm, and $\Delta = 0.3\%$ to 0.8%, where $\Delta = (n_1 - n_2)/n_1$. When separation c and the relative refractive-index difference Δ take larger values, the value of L tends to be large.

10.1.4 Examples of Directional Couplers

The bulk-type coupler is not very useful for single-mode fibers because of its large insertion loss. It is also large in size. Fiber-type directional couplers are classified into two types: a polished-fiber type and a fused-fiber type. In the polished-fiber type, couplers are fabricated by polishing each of two fibers. The polished part is different from the endface of fibers, which is the case for optical connectors. In the case of fabricating couplers, a side of two fibers is polished. A coupler is made by placing the two polished faces in contact, and index-matching materials are used at the contact surfaces. The drawbacks of this type are that the polishing must be done accurately and that the fabricated couplers have unstable characteristics in several environments. Therefore, the fused-fiber type and the waveguide type are preferable, especially for systems using single-mode fibers.

Fused-Fiber Type

The structure and fabrication methods of fusing fibers are shown in Figure 10.5. Fibers are heated by a flame or by an electric discharge and are simultaneously pulled (elongation). In this process, fibers are fused together. The softened parts are formed into a tapered shape. In the tapered part, the distance between cores in fibers becomes close and nonnegligible coupling takes place between the cores. This coupling realizes the directional coupler. Although a 3-dB coupler is common, couplers with other coupling ratios, such as a 1:10 ratio, are also made. To make couplers that have accurate coupling ratios, optical power is launched into a fiber and the output power from a fiber is monitored during the heating and pulling process. Since fused-fiber couplers are made from an ordinary fiber, the coupling efficiency between fibers and fused-fiber couplers is very high. The typical characteristics of fused-fiber-type directional couplers are listed in Table 10.1 in the case of applications using single-mode fiber. The excess loss, coupling loss, and isolation characteristics are better than those of waveguide-type directional couplers with the current technology. The first drawback of fused-fiber-type directional couplers is their noncompactness when they are used for a large port coupler by being cascaded, and the second drawback is the lack of integration with other optical devices. The size of a fused-fiber-type directional coupler is smaller than that of a bulk-type coupler, and is nearly the same size as a waveguide-type coupler. However, fused-fiber-type devices are large when they are used in a cascade as a unit device to make star couplers such as a 2 × 8 star coupler.

As with mass-fusion splices, simultaneous fabrication of fiber-type directional couplers has been proposed and demonstrated using a fiber ribbon [3]. Using a four-fiber ribbon, four fused-fiber-type directional couplers are fabricated simultaneously using two burners arranged on the opposite side [3].

In ordinary applications, a wavelength-independent directional coupler is desirable. Wavelength-flattened or wide-bandwidth directional couplers by using two fibers with a slightly different propagation constant have been proposed [4]. Several

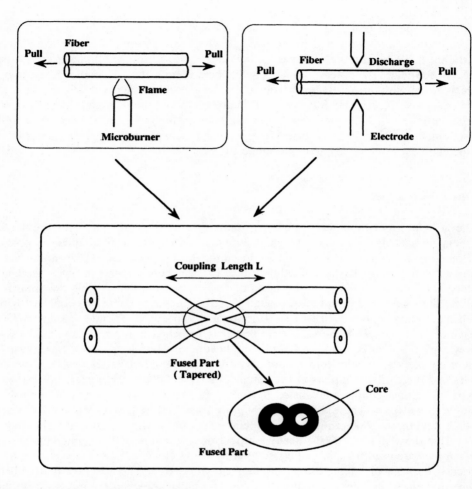

Figure 10.5 Fiber-type directional coupler.

Table 10.1

Directional Coupler for Single-Mode Fibers

	Excess Loss (dB)	Coupling Loss to a Single-Mode Fiber (dB)	Isolation (dB)
Fused-fiber coupler	0.1–0.2	Fusion splice: <0.1 ———————————— Connector: <0.3	40–50
Waveguide-type coupler (silica-based planar waveguide)	0.3–0.5	0.3–1*	30–40

*Depends on core size and Δ.

methods have been proposed for obtaining different propagation constants, such as (1) the use of fibers with different diameters, (2) the use of fibers with different profiles, or (3) using different tapering ratios for two identical fibers [4]. In the case of (3), a fiber is heated and elongated in advance (pretaper) to obtain nonequal dimensions, and then the pretapered fiber and an ordinary fiber are heated together and elongated [4]. In this method, two fibers are originally the same fiber. The fabricated coupler has 50% ± 9% coupling over the wavelength range from 1.23 to 1.57 μm. Wavelength-flattened directional couplers have been fabricated with another approach using two fibers with different fiber parameters [5]. In this case, two fibers of different type are heated and elongated by the ordinary fabrication process.

Polarization-maintaining fiber couplers have been demonstrated for special applications, and some of them are explained here [6–10]. They are made from polarization-maintaining fibers, such as a PANDA fiber [6–9] and an elliptic-jacket fiber [10]. Structures of PANDA and elliptic-jacket fibers are shown in Figure 1.8(d,c), respectively. For the case of using PANDA fibers, a polarization-maintaining fiber coupler is shown in Figure 10.6(a). The coupler is fabricated by the method used for fabricating ordinary fused-fiber couplers [9]. The angular alignment for two parallel PANDA fibers must be carefully made during the fabrication process. Two types of polarization-maintaining fiber couplers are shown in Figure 10.6(b,c). The polarization direction of an input light is preserved at two output ports in the first type (Fig. 10.6(b)). In this coupler, x-polarized lights are output from two output ports when an x-polarized light is input, while y-polarized lights are output in the case of the y-polarized input light. In the second type of coupler (Fig. 10.6(c)), an x-polarized light is output from only one output port when an x-polarized light is input, while a y-polarized light is only output from another output port in the case of a y-polarized input light. Two orthogonal polarization states are separated at the output ports for an input light with x- and y- polarized components, and this type has been proposed for an application of a polarizing beam splitter [7]. These characteristics are generally wavelength-dependent.

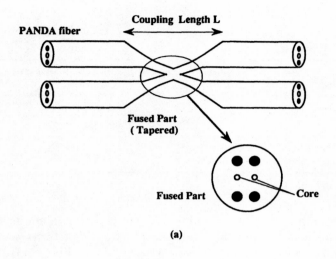

(a)

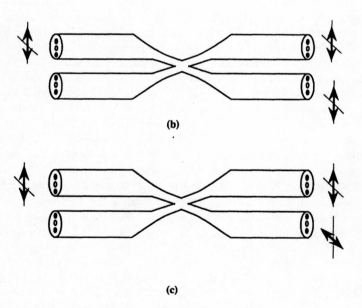

(b)

(c)

Figure 10.6 Polarization-maintaining fiber coupler. (After [9].)

Waveguide Type

Waveguide-type optical circuits (optical integrated circuits) were proposed in 1969 [11]. Since then, many optical glass waveguides have been investigated. They are fabricated by ion exchange, ion plantation, chemical vapor deposition (CVD), and so forth. One of the early works used the sputtering technique to make a glass waveguide [12]. Recently, silica planar waveguides have been proposed [13,14] and they may be promising, because the composition is similar to that of optical fibers and they have the potential of low-loss waveguides. Optical waveguides or circuits are formed on a substrate by a combination of the processes of glass film deposition and etching. Several deposition processes, such as flame hydrolysis deposition [14], CVD [15,16], and electron-beam vapor deposition [17], have been proposed and used. Figure 10.7 shows the fabrication process using flame hydrolysis deposition and reactive ion etching as an example of a waveguide fabrication process [14,18]. The waveguide has a SiO_2-TiO_2 core. The measured propagation loss is reported to be 0.10 dB/cm at 1.32 μm [14]. A very-low-loss waveguide is fabricated by using a SiO_2-GeO_2 core, and the measured loss, including bend loss, is 0.04 dB/cm at 1.55 μm and the estimated loss in the straight waveguide part is $\pm$0.01 dB/cm [19].

A directional coupler can be formed in a silica planar waveguide. Several approaches for realizing wavelength-flattened or wide-bandwidth directional couplers have been proposed. One approach, which uses a Mach-Zehnder interferometer, is shown in Figure 10.8 [20]. The Mach-Zehnder interferometer uses two ordinary directional couplers, as shown in the figure. The fabricated coupler has the following parameters: 8 × 8 μm core (rectangular cores), $\Delta = 0.25\%$, separation $c = 7$ μm, and a small optical path difference $\Delta L = 0.6$ μm between the two waveguide arms indicated in the figure. It is pointed out in [20] that the wavelength-flattened couplers with this structure result both from the inequality of the two couplers used in the Mach-Zehnder interferometer and from the waves transmitting in two arms with out of phase. The small coupling ratio variation (20% $\pm$ 1.9%) has been achieved for the wavelength range of 1.25 to 1.65 μm. Another approach uses an ordinary coupler configuration with an asymmetric structure [21]. The configuration is shown in Figure 10.9. The rectangular cores of 8 × 8 μm and 6 × 8 μm for the two arms, $\Delta = 0.25\%$, and separation $c = 3$ μm are used in [21], and a coupling ratio variation of 50% $\pm$ 5% for the 1.2- to 1.6-μm wavelength range has been obtained.

Typical characteristics of waveguide-type directional couplers are listed in Table 10.1. The characteristics of excess loss, coupling loss, and isolation of waveguide-type directional couplers with the current technology are worse than those of fiber-type directional couplers. In the future, as waveguide-type devices are intensively investigated, their characteristics will be seen as similar to those of fiber-type couplers.

Optical planar waveguides other than the silica planar waveguides mentioned above have been proposed and fabricated, and some of them are explained below.

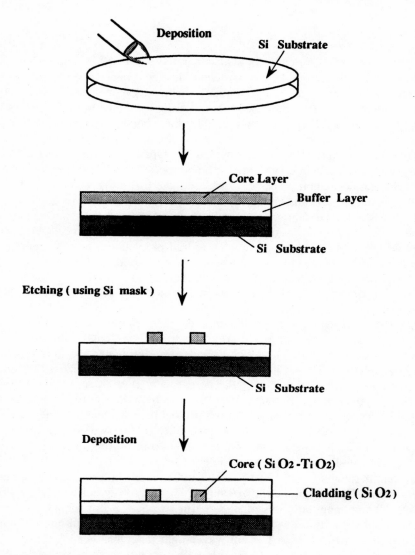

Figure 10.7 Waveguide-type directional coupler. (After [18].)

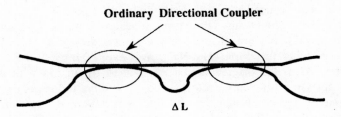

Figure 10.8 Mach-Zehnder interferometer–type directional coupler.

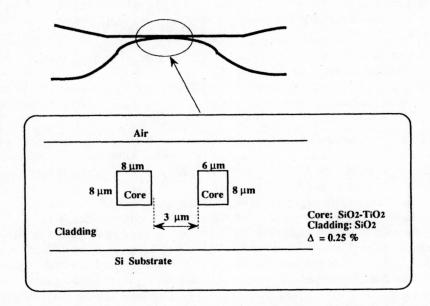

Figure 10.9 Waveguide-type directional coupler with asymmetric structure. (After [21])

These are and will be used not only for a directional coupler but also for several other passive optical components. Glass waveguides made by the ion exchange method are reviewed in [22]. An ion diffuses into the glass networks and a certain ion is exchanged with the diffused ion. The Na^+ ion is exchange by Tl^+, K^+, or Ag^+ ions in most cases. When the original ion is exchanged by an ion with a larger ionic radius or by an ion with a larger electric polarizability, the refractive index of an exchanged part increases. When the situation is reversed, the refractive index decreases. For example, a Tl^+ ion has an ionic radius that is 1.57 times larger than that of a Na^+ ion, and an electric polarizability that is 12.7 times larger. In this case, the exchanged part has a higher refractive index. The ion exchange method is a low-temperature process of about 300° to 600°C. Passive optical components have been made by photolithography with a ion exchange in glass [23]. The proposed process takes three steps:

1. A mask material is deposited on a starting glass. The mask acts as a barrier to ion diffusion at the third step.
2. Optical circuit patterns are written (etched) into the deposited mask by a photolithographic process.
3. An ion exchange is made to form a waveguide.

The mask material candidates titanium, aluminum, SiO_2, or combinations of these have been reported. In the third step, thallium has been chosen as the diffusion ion in [23]. Excess loss of the fabricated 1×2 coupler with a 50-μm core has been reported to be under 0.2 dB [23].

Antiresonant reflecting optical waveguides, which are silica-based single-mode waveguides, have been proposed for a photonic integration circuit [24–26]. The structures are shown in Figure 10.10; excitation of these waveguides is shown in Figure 10.10(a). The structures have two cladding layers and are divided into two types. The first type is called arrow, and its first cladding layer has a higher refractive index than that of the core layer, and its second cladding layer has the same refractive index as that of the core layer [24]. The first cladding layer is polarization-dependent and it may be useful for realizing some functional components. The second type of structure is called arrow-B, and its first cladding layer has a lower refractive index than that of the core layer, and its second cladding layer has the same refractive index [25]. This is intended to realize a waveguide that is not polarization-dependent, and it is useful for ordinary applications. Unlike ordinary waveguides, these structures indicate that they are leaky waveguides. Waves are guided by the total internal reflection at the core-cladding interfaces in ordinary waveguides; on the other hand, the antiresonant reflecting optical waveguides use the antiresonant reflection for wave guidance. The layers of the antiresonant reflecting waveguides form a Fabry-Perot resonator, which has resonances and antiresonances. The antiresonances are spectrally broad, while the resonances are spectrally narrow (see Section 11.2). The antiresonant reflecting optical waveguides have been proposed in order to have the

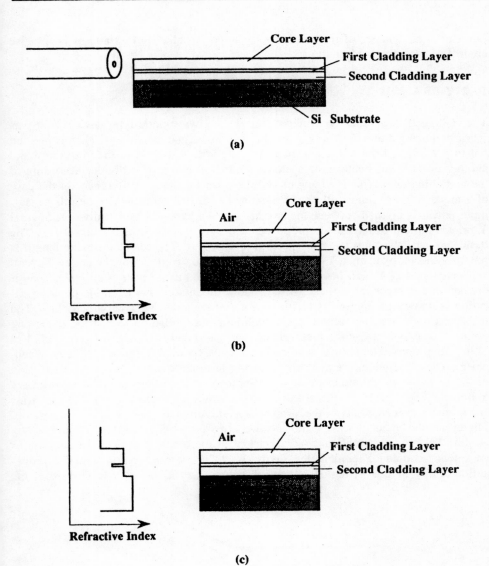

Figure 10.10 Antiresonant reflecting optical waveguide: (a) excitation by a fiber; (b) arrow; (c) arrow-
B. (After [24–26].)

fabrication advantages of thin cladding to suppress radiation loss and of no precise control of the waveguide index.

10.2 Y-BRANCH

The Y-branch is another type of coupler. It is a waveguide-type device, and was fabricated using a thin-film waveguide in early investigation stages [27]. It is pointed out in [27,28] that the Y-branch, in which one arm is straight to the initial arm and another is inclined, behaves as a mode splitter or as a power divider according to the branching angle (inclined angle) and the refractive-index difference. In the case of a mode splitter, almost all power remains in the straight arm. On the other hand, input power is divided between the two arms in the case of a power divider. Several Y-branch structures have been proposed, and they are shown in Figure 10.11. The basic ordinary structure is shown in Figure 10.11(a). The realized device length is possibly shorter by using a Y-branch when compared to a directional coupler. Several applications using Y-branches have been proposed and used, such as an optical power divider, a combiner, an optical switch, and a modulator. An example of an application is shown in Figure 10.11(b): an interferometer using two Y-branches. The drawback of Y-branches is the loss, which is mainly radiation loss at the branching point. A way to decrease the loss has been proposed and is shown in Figure 10.11(c) [29]. The proposed construction is called antenna-coupled Y-branch, after its mechanism. The waveguides are assumed to be single-mode waveguides. The explanation of the low loss in an antenna-coupled Y-branch is as follows [29]. In an ordinary Y-branch (Fig. 10.11(a)), the phase fronts of waves near the input of the two arms are almost perpendicular to the propagation direction of the incident waveguide. Therefore, the incident waves into the two arms are oblique to the two arm waveguides. This causes significant radiation loss at the branching point. In the case of an antenna-coupled Y-branch, the fields spread into the n_3 region, and the propagation ability of this n_3 region is weak because $n_3 < n_1$ ($n_3 < n_2$). Therefore, the

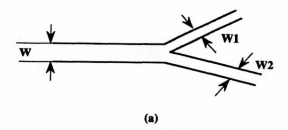

(a)

Figure 10.11 Y-branch:(a) one example of Y-branch; (b) Y-branch interferometer; (c) antenna-coupled Y-branch. (After [29].)

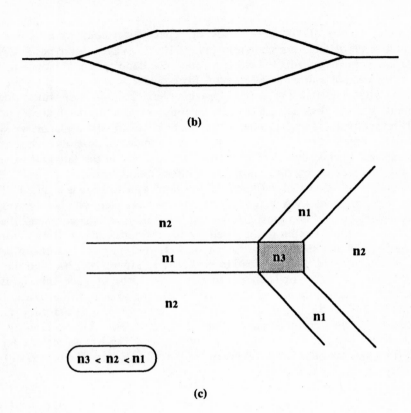

(b)

(c)

Figure 10.11 (Continued)

field in the initial waveguide radiates and spreads into the n_3 region. As a result of the field spreading, the phase fronts are perpendicular to the two arm waveguides. The radiation loss is small when compared to that in the ordinary Y-branch. Experimentally, the branching losses of 3.4 and 1.0 dB have been reported for the ordinary and the antenna-coupled Y-branches, respectively [29].

10.3 STAR COUPLER

A star coupler is used for mixing and branching lightwaves. As the power is branched in a star coupler, intrinsic branching loss (or splitting loss) L_B exists. It is expressed as follows in the case of equal division:

$$L_B = -10 \log(1/N) \quad \text{(dB)} \tag{10.42}$$

For example, $L_B = 9$ dB for $N = 8$. Several types of star couplers have been proposed, and some of them are shown in Figure 10.12. Waveguide-type $2 \times N$ or $1 \times N$ star couplers are constructed using a directional coupler or a Y-branch as a unit block. In the case of a directional coupler, fiber-type $2 \times N$ star couplers are made by fusion splicing of unit 2×2 fiber-type couplers. Examples of waveguide-type single-mode star couplers using a directional coupler and a Y-branch are reported in [30,31], respectively. These examples are made by using silica planar waveguide technologies. Any port of a $2 \times N$ or $1 \times N$ star coupler can possibly be constructed with the cascading unit coupler blocks. However, the size of the fabricated star couplers using this method becomes large for a larger number of N.

Direct construction without using a unit block is also shown in Figure 10.12. $1 \times N$, $2 \times N$, or generally $N \times N$ star couplers are fabricated with both waveguide-type and fiber-type technologies. In the case of fiber-type star couplers, they are made with the method similar to that used for fused-fiber couplers. Both multimode fiber-type [32,33] and single-mode fiber-type [34] star couplers have been demonstrated. An example of their fabrication is shown in Figure 10.13(a) and the cross section of the elongated part is shown in Figure 10.13(b). A glass tube is used to avoid breakage, deformation, and excess-heating at the stretched-fiber portion [33]. The refractive index of the glass tube is selected to be lower than that of a fiber, so that strong mode field confinement in cores is realized [34]. All star couplers have an excess loss L_E in addition to the branching loss L_B. The excess loss L_E for $N \times N$ fused-fiber star couplers using GI fibers with a 50-μm core is reported to be approximately expressed as [33]

$$L_E \approx \log N \tag{10.43}$$

This equation is derived from the measured excess losses for fabricated star couplers from $N = 2$ to $N = 100$. For 100×100 star couplers, excess loss $L_E = \log 100 = 2$ dB.

In the case of waveguide-type single-mode star couplers, formation of a slab waveguide region, which connects input and output channel waveguide arrays, has been proposed [35,36]. The formed slab waveguide region, shown in Figure 10.12, works just like a free space in radio systems and serves as a lightwave mixer. A similar configuration has been proposed and fabricated in a bulk-type star coupler for multimode fibers at the early investigation stage of star couplers [37]. The input power from any of the N input channels radiates into the slab waveguide region, and the radiated power is received by the output array. Uniform branching or splitting among N output channels is realized when the uniform radiation pattern is produced over the output channel waveguide arrays. Some 19×19 star couplers have been fabricated on a silicon substrate, and the reported excess loss L_E is about 3.5 dB, which is not very large when compared to the intrinsic branching loss $L_B = 12.8$ dB ($L_B = 10 \log 19$) [35]. The excess loss L_E is reported to decrease to about 1.5 dB,

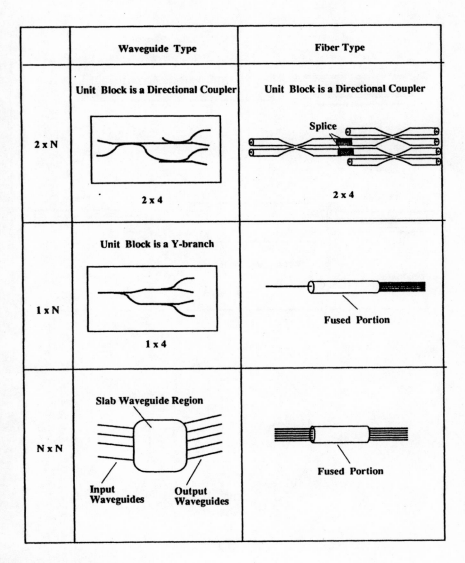

	Waveguide Type	Fiber Type
2 x N	**Unit Block is a Directional Coupler** 2 x 4	**Unit Block is a Directional Coupler** Splice 2 x 4
1 x N	**Unit Block is a Y-branch** 1 x 4	Fused Portion
N x N	Slab Waveguide Region Input Waveguides Output Waveguides	Fused Portion

Figure 10.12 Star couplers.

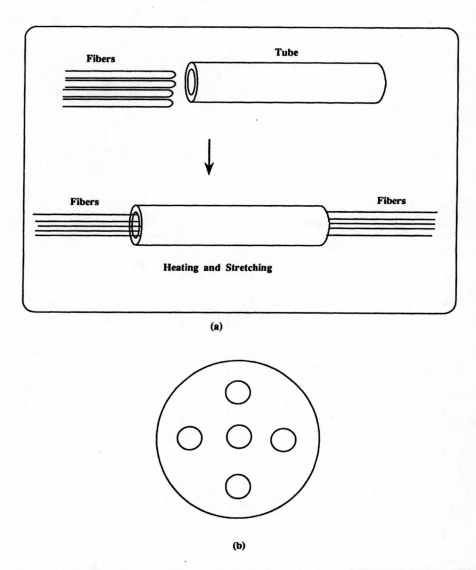

(a)

(b)

Figure 10.13 Fabrication of fiber-type star couplers: (a) fabrication method; (b) cross section of elongated part. (After [33,34].)

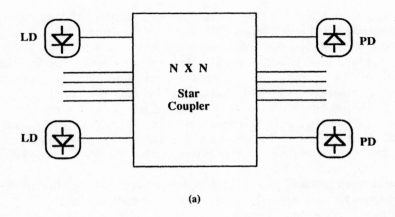

(a)

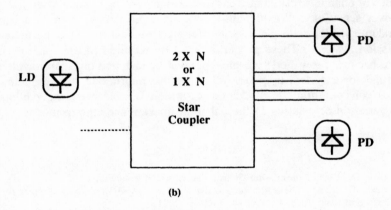

(b)

Figure 10.14 Application of star coupler: (a) LAN; (b) PON.

which is the case for 8 × 8 star couplers using silica planar waveguide technologies, from precise design and optimization [36]. The reported device size is 5 × 26 mm for an 8 × 8 star coupler, and the waveguide spacing between input (or output) waveguides at both ends is 0.25 mm [36]. This reported device size indicates the compactness of this approach. Wavelength-insensitive light splitting has been realized by properly designing the mode coupling at the input channel arrays [38]. For this purpose, the nonuniform core in an input channel waveguide is used. Wavelength-insensitive light splitting has been obtained with an excess loss $L_E = 1.5$ dB for a wavelength range 1.3 to 1.55 μm by optimizing the tapering position and the taper ratio [38]. The polarization dependence of average excess loss is reported to be smaller than 0.3 dB [38].

Several applications of star couplers has been proposed, and two of them are shown in Figure 10.14: LAN in local communication applications and PON in telecommunication applications. Star couplers can mix input optical signals, and the mixed signals are transmitted to any output channel. Therefore, star couplers are considered to be a free space in radio communications, and they introduce the merits of radio transmission systems into fiber-optic transmission systems. However, they also bring the disadvantages of radio transmission systems simultaneously. The interference of channels results in cross talk of neighboring channels and a degradation of receiver sensitivity. Wavelength-independent properties, such as a wavelength-independent splitting ratio, are required for star couplers when we use them in WDM transmission systems. These properties may be required in the case of non-WDM systems, because laser diode wavelengths will be changed or other wavelengths will be overlaid for a system upgrade. When we use multimode star couplers, we must take into consideration the modal noise, which is caused by the random speckle pattern variation originating in the coherent interference among modes.

REFERENCES

[1] Marcuse, D., Light Transmission Optics, Van Nostrand Reinhold, 1972.

[2] Marcatili, E. A. J., "Dielectric Rectangular Waveguide and Directional Coupler for Integrated Optics," Bell Syst. Tech. J., Vol. 48, 1969, p. 2071.

[3] Yokohama, I., K. Chida, H. Hanafusa, and J. Noda, "Novel Mass-Fabrication of Fiber Couplers Using Arrayed Fiber Ribbons," Electron. Lett., Vol. 24, 1988, p. 1147.

[4] Mortimore, D. B., "Wavelength-Flattened Fused Couplers," Electron. Lett., Vol. 21, 1985, p. 742.

[5] Hanafusa, H., Y. Takeuchi, and J. Noda, "Wavelength Flattened Optical Fiber Coupler II," IEICE of Japan, Natl. Conf. Rec. on Commun., No. C-208, 1989 (in Japanese).

[6] Kawachi, M., B. S. Kawasaki, K. O. Hill, and T. Edahiro, "Fabrication of Single-Polarisation Single-Mode-Fiber Couplers," Electron. Lett., Vol. 18, 1982, p. 962.

[7] Yokohama, I., K. Okamoto, and J. Noda, "Fiber-Optic Polarising Beam Splitter Employing Birefringent-Fiber Coupler," Electron. Lett., Vol. 21, 1985, p. 415.

[8] Yokohama, I., M. Kawachi, K. Okamoto, and J. Noda, "Polarisation-Maintaining Fiber Couplers With Low Excess Loss," Electron. Lett., Vol. 22, 1986, p. 929.

[9] Yokohama, I., J. Noda, and K. Okamoto, "Fiber-Coupler Fabrication With Automatic Fusion-

Elongation Processes for Low Excess Loss and High Coupling-Ratio Accuracy," IEEE J. Lightwave Technol., Vol. LT-5, 1987, p. 910.

[10] Abebe, M., C. A. Villarruel, and W. K. Burns, "Reproducible Fabrication Method for Polarization Preserving Single-Mode Fiber Couplers," IEEE J. Lightwave Technol., Vol. 6, 1988, p. 1191.

[11] Miller, S. E., "Integrated Optics: An Introduction," Bell Syst. Tech. J., Vol. 48, 1969, p. 2059.

[12] Goell, J. E., and R. D. Standley, "Sputtered Glass Waveguide for Integrated Optical Circuits," Bell Syst. Tech. J., Vol. 48, 1969, p. 3445.

[13] Izawa, T., H. Mori, Y. Murakami, and N. Shimizu, "Deposited Silica Waveguide for Integrated Optical Circuits," Appl. Phys. Lett., Vol. 38, 1981, p. 483.

[14] Kawachi, M., M. Yasu, and T. Edahirio, "Fabrication of S_iO_2-T_iO_2 Glass Planar Optical Waveguides by Flame Hydrolysis Deposition," Electron. Lett., Vol. 19, 1983, p. 583.

[15] Valette, S., "State of the Art of Integrated Optics Technology at LETI for Achieving Passive Optical Components," J. Modern Opt., Vol. 35, 1988, p. 993.

[16] Henry, C. H., G. E. Blonder, and R. F. Kazarinov, "Glass Waveguides on Silicon for Hybrid Optical Packaging," IEEE J. Lightwave Technol., Vol. 7, 1989, p. 1530.

[17] Imoto, K., T. Asai, H. Sano, and M. Maeda, "Silica Glass Waveguide Structure and Its Application to a Multi/Demultiplexer," European Conf. on Optical Communications (ECOC'88), 1988, p. 577.

[18] Takato, N., K. Jinguji, M. Yasu, H. Toba, and M. Kawachi, "Silica-Based Single-Mode Waveguides on Silicon and Their Application to Guided-Wave Optical Interferometers," IEEE J. Lightwave Technol., Vol. 6, 1988, p. 1003.

[19] Kominato, T., Y. Ohmori, H. Okazaki, and M. Yasu, "Very Low-Loss GeO_2-Doped Silica Waveguides Fabricated by Flame Hydrolysis Deposition Method," Electron. Lett., Vol. 26, 1990, p. 327.

[20] Jinguji, K., N. Takato, A. Sugita, and M. Kawachi, "Mach-Zehnder Interferometer Type Optical Waveguide Coupler With Wavelength-Flattened Coupling Ratio," Electron. Lett., Vol. 26, 1990, p. 1326.

[21] Takagi, A., K. Jinguji, and M. Kawachi, "Broadband Silica-Based Optical Waveguide Coupler With Asymmetric Structure," Electron. Lett., Vol. 26, 1990, p. 132.

[22] Findakly, T., "Glass Waveguides by Ion Exchange: A Review," Opt. Eng., Vol. 24, 1985, p. 244.

[23] Beguin, A., T. Dumas, M. J. Hackert, R. Jansen, and C. Nissim, "Fabrication and Performance of Low Loss Optical Components Made by Ion Exchange in Glass," IEEE J. Lightwave Technol., Vol. 6, 1988, p. 1483.

[24] Duguay, M. A., Y. Kokubun, T. L. Koch, and L. Pfeiffer, "Antiresonant Reflecting Optical Waveguides in SiO_2-Si Multilayer Structures," Appl. Phys., Lett., Vol. 49, 1986, p. 13.

[25] Baba, T., and Y. Kokubun, "New Polarization-Insensitive Antiresonant Reflecting Optical Waveguide (ARROW-B)," IEEE Photon. Technol. Lett., Vol. 1, 1989, p. 232.

[26] Sakamoto, T., and Y. Kokubun, "Monolithic Integration of Photodetector and ARROW-Type Interferometer for Detecting Phase Difference Between Two Optical Paths," Japanese J. of Appl. Phys., Vol. 29, 1990, p. L96.

[27] Yajima, H., "Dielectric Thin-Film Optical Branching Waveguide," Appl. Phys. Lett., Vol. 22, 1973, p. 647.

[28] Burns, W. K., and A. F. Milton," Mode Conversion in Planar-Dielectric Separating Waveguides," IEEE J. Quantum Electron., Vol. QE-11, 1975, p. 32.

[29] Hanaizumi, O., M. Miyagi, M. Minakata, and S. Kawakami, "Antenna Coupled Y Junctions in 3-Dimensional Dielectric Optical Waveguides," European Conf. on Optical Communications (ECOC'85), 1985, p. 179.

[30] Okuno, M., A. Takagi, M. Yasu, and N. Takato, "2 × 8 Single-Mode Guided-Wave Star Coupler," IEICE of Japan, Natl. Conf. Rec. on Commun., No. C-501, Spring 1989 (in Japanese).

[31] Kobayashi, S., T. Kito, Y. Hida, and M. Yamaguchi, "Optical Waveguide 1 × 8 Splitter Module," NTT R&D, Vol. 39, No. 6,1990, p. 931 (in Japanese).

[32] Rawson, E. G., and M. D. Bailey, "Bitaper Star Couplers With up to 100 Fiber Channels," Electron. Lett., Vol. 15, 1979, p. 432.

[33] Imoto, K., M. Maeda, H. Kunugiyama, and T. Shiota, "New Biconically Tapered Fiber Star Coupler Fabricated by Indirect Heating Method," IEEE J. Lightwave Technol., Vol. LT-5, 1987, p. 694.

[34] Mortimore, D. B., and J. W. Arkwright, "Monolithic Wavelength-Flattened 1 × 7 Single-Mode Fused Coupler," Electron. Lett., Vol. 25, 1989, p. 606.

[35] Dragone, C., C. H. Henry, I. P. Kaminow, and R. C. Kistler, "Efficient Multichannel Integrated Optics Star Coupler on Silicon," IEEE Photon. Technol. Lett., Vol. 1, 1989, p. 241.

[36] Okamoto, K., H. Takahashi, S. Suzuki, A. Sugita, and Y. Ohmori, "Design and Fabrication of Integrated-Optic 8 × 8 Star Coupler," Electron. Lett., Vol. 27, 1991, p. 774.

[37] Nosu, K., and R. Watanabe, "Slab Waveguide Star Coupler for Multimode Optical Fibers," Electron. Lett., Vol. 16, 1980, p. 608.

[38] Okamoto, K., H. Takahashi, M. Yasu, and Y. Hibino, "Fabrication of Wavelength-Insensitive 8 × 8 Star Coupler," IEEE Photon. Technol. Lett., Vol. 4, 1992, p. 61.

Chapter 11
Optical Filters

Several types of optical filters have been and will continue to be used in optical transmission systems. WDM and OFDM systems use WDM and OFDM filters, respectively. These filters are used for multiplexing or demultiplexing lightwaves, just like electric filters in frequency-division multiplexing (FDM) systems. Both WDM and OFDM systems are wavelength division multiplexing systems, and the difference is the density in the wavelength or the frequency domain; that is, WDM is coarse, while OFDM is fine. The wavelength spacing $\Delta\lambda$ of ordinary WDM systems is about 100 to 300 nm, and that of OFDM systems is about 0.08 nm when $\Delta f = 10$ GHz. Therefore, characteristics of WDM filters are discussed in terms of wavelength (nm), while those of OFDM filters can be discussed in a term of frequency (GHz).

There are many types of electric filters, such as the low-pass filter (LPF), bandpass filter (BPF), band-rejection filter (BRF), and transversal filter. Because of the progress in the fabrication technology of optical waveguides and photonic ICs, several optical filters have actually been fabricated and evaluated in addition to WDM and OFDM filters. Many fabricated optical filters are analogous to existing electrical filters in microwave integrated circuits (MIC). MIC waveguides such as strip lines correspond to optical waveguides. The difference between them is that metals are used for guiding waves in MICs, while dielectric materials with a high refractive index are used for guiding waves in optical waveguides. The circuits in MICs are designed by treating electrical waves rather than a current and a voltage which are commonly used at low-frequency IC. Therefore, there may be a natural analogy between MICs and photonic ICs. In this chapter, WDM filters, OFDM filters, BRFs, including the Bragg reflector, and optical signal processing filters, such as the optical transversal filter, are explained.

11.1 WDM FILTER

11.1.1 Bulk-Type WDM Filter

Bulk-type WDM filters frequently use dispersive devices, such as a grating or a prism. The dispersive device transforms the wavelength difference to the light propagation direction difference. Some types of WDM filters use dielectric thin-film filters, shown in Figure 11.1. In the case of a WDM filter using a grating, the principle of wavelength-division multiplexer or demultiplexer is explained below. The light-focusing position difference of each wavelength to the reference position is expressed by the light propagation direction difference; that is,

$$\Delta x_1 = \left(\frac{d\theta}{d\lambda}\right) \Delta\lambda_1 L \tag{11.1}$$

$$\Delta x_2 = \left(\frac{d\theta}{d\lambda}\right) \Delta\lambda_2 L \tag{11.2}$$

and so on

$$\Delta\lambda_1 = \lambda_1 - \lambda_2 \tag{11.3}$$

$$\Delta\lambda_2 = \lambda_2 - \lambda_3 \tag{11.4}$$

......

where θ is the diffracted angle and L is the length shown in Figure 11.1(a). The symbols Δx_i and $\Delta\lambda_i$ ($i = 1, 2, \ldots$) are the position and the wavelength difference, respectively. The angular dispersion, which is an important parameter, is given as [1]

$$\frac{d\theta}{d\lambda} = \frac{m}{\Lambda \cos\theta} \tag{11.5}$$

where m is an integer and Λ is the grating constant (the pitch of a grating). To obtain large angular dispersion, Λ must be small. A similar principle holds for a WDM filter using a prism.

Optical filters using multilayer dielectric thin films are well known in optics. This type of filter uses the interference phenomenon of optical waves [1]. The principle of multilayer dielectric thin films is explained in Figure 11.2. The thin film is assumed to have a thickness of $\lambda/4$ (a quarter wavelength). The characteristic impedance of the film is expressed as [2]

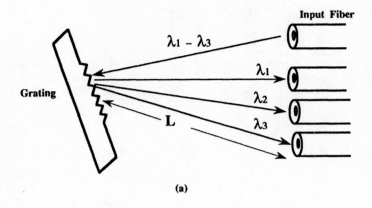

(a)

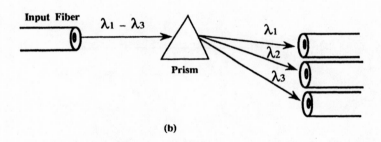

(b)

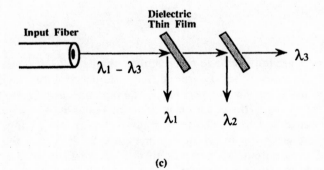

(c)

Figure 11.1 Bulk-type WDM filter: (a) grating; (b) prism; (c) dielectric thin film.

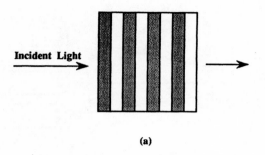

(a)

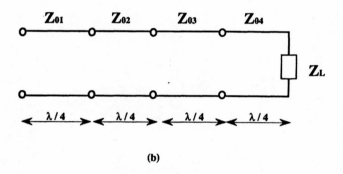

(b)

Figure 11.2 Dielectric thin-film WDM filter: (a) multilayer thin film; (b) equivalent electric circuit.

$$Z_0 = \sqrt{\frac{\mu}{\varepsilon}} = \frac{1}{n} \sqrt{\frac{\mu_0}{\varepsilon_0}} \qquad (11.6)$$

where μ, ε, and n are the magnetic permeability, the permittivity, and the refractive index, respectively, and μ_0 and ε_0 are the magnetic permeability for a vacuum and the permittivity for a vacuum, respectively. The equation $\mu = \mu_0$ holds in the case of an optical frequency. This equation is used for obtaining (11.6) with $\varepsilon = n^2 \varepsilon_0$. Multilayer dielectric thin films, shown in Figure 11.2(a), are analogous to the cascaded circuits (transmission lines) shown in Figure 11.2(b), and the characteristic impedance Z_0 of the circuit corresponds to that of the film. It is known that the various frequency response can be obtained with a suitable design of the characteristic impedances Z_{0n} in cascaded quarter-wavelength circuits [2]. Analogously, several types of optical filters can be obtained with a suitable design of the characteristic impedances Z_{0n}, which corresponds to the refractive indexes n_n in this case. For example, impedance matching (no-reflection) for a line with Z_{01} is realized by using the quarter-wavelength circuits with Z_{02} when the equation $Z_{01}Z_L = Z_{02}^2$ holds, where Z_L is the load impedance. In the case of optical thin films, the no-reflection property is analogically obtained for perpendicular incident optical waves when $n_1 n_3 = n_2^2$ (n_3 corresponds to Z_L) holds. The dielectric multilayer thin films are commonly fabricated with a deposition of several coating materials forming in multilayers. The selection of coating materials is very important to obtain good optical and mechanical properties. The fabricated thin-film filters for use in the 0.8- and 1.3-μm wavelength range are explained here as an example [3]. In [3], electron-beam heating was adopted for the deposition of coating materials. TiO_2 and ZrO_2 were used as high-index materials, and SiO_2 and Al_2O_3 as low-index materials. To obtain good BPFs, a 23-layer thin film was fabricated [3].

The filters mentioned in Figure 11.1 are shown only to explain their principles. The actual filter devices, such as wavelength-division multiplexers and demultiplexers, are implemented in other configurations to decrease size and insertion loss, especially in the case of single-mode fiber applications. Bulk-type WDM filters have been demonstrated with a micro-optical design or with the aid of a planar waveguide design [4–8]; they are shown in Figures 11.3 and 11.4. Rod lenses are frequently used in a micro-optical design for focusing light. There is a brief explanation of the rod lens at the end of this section. WDM filters using a grating are shown in Figure 11.3. A WDM filter for single-mode fibers in two-way operation with 2 × 2 channels has been reported in [4], and one of the proposed structures is shown in Figure 11.3(a). Four wavelengths, namely 1.05/1.15 μm and 1.3/1.5 μm in this case, are mixed and separated by the thin-film filter and the grating. The reported passband width, insertion loss, and cross talk between channels are abut 3 nm, 2 to 3 dB, and lower than −40 dB, respectively [4]. A 20-channel demultiplexer in the 1.1- to 1.6-μm wavelength region for a single-mode fiber has been reported in [5], and it is shown in Figure 11.3(b). Since this is designed for a demultiplexer, the receiving fibers are SI multimode fibers with a 40-μm core diameter, which are connected to detectors, while an input fiber is a single-mode fiber. The reported insertion losses for 20 channels are between 1.9 and 3.5 dB. The channel spacing ranges from 27 to 31 nm [5]. Examples of WDM devices using dielectric thin-film filters are shown in Figure 11.4 [6,7]. These examples are a BPF type. In the case of Figure 11.4(a), six input wavelengths λ_1 to λ_6 are separated by the thin-film filters. The reported insertion loss is about 1 dB and six wavelengths are located in the 0.7- to 1.0-μm wavelength region. Figure 11.4(b) schematically shows another type of device designed for a multimode fiber, which uses a planar waveguide [7]. Since dielectric thin-film filter chips and planar waveguides are used in this device, it is a hybrid type using bulk and a waveguide types. The planar waveguide is used both as a waveguide and as an alignment part, the latter designed for aligning fibers and a thin-film filter chip, as shown in Figure 11.4(b) [7]. Waves with 1.2- and 1.3-μm wavelengths are designed for transmitting, and waves with 0.81 and 0.89 μm for reflecting in the reported device, where the configuration is more complex than that

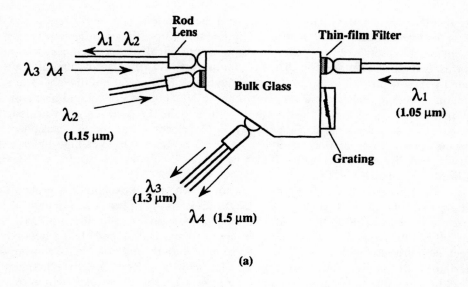

(a)

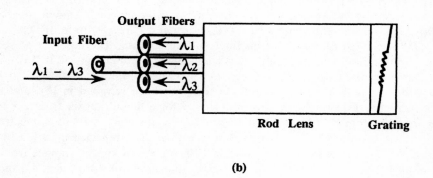

(b)

Figure 11.3 Bulk-type WDM device using grating filter. (After [4,5].)

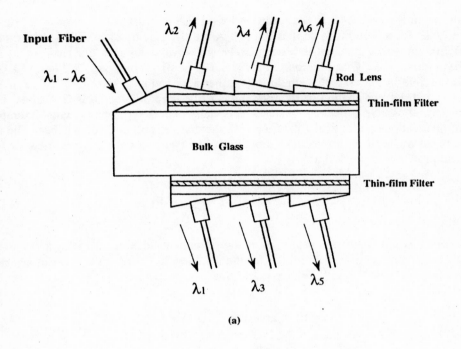

(a)

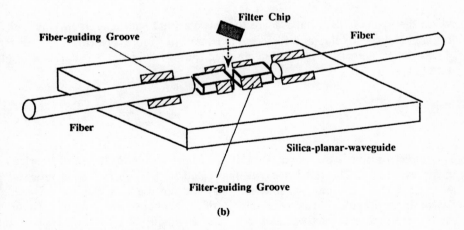

(b)

Figure 11.4 Bulk-type WDM device using thin-film filter. (After [6,7].)

shown in Figure 11.4(b). The insertion loss is 3.5 to 5 dB with crosstalk of -30 to -50 dB for a multimode application [7]. A multi/demultiplexer with a similar configuration, using gradient-index ion-exchange waveguides in a glass substrate, has been reported for GI multimode fibers [8]. The configuration used in Figure 11.4(b) is possible in the case of a single-mode fiber application.

A brief explanation of the rod lens is made here, because it is frequently used in several passive optical components, especially in bulk-type components. Examples of applications are optical filters, optical attenuators, and optical switches. The rod lens has a graded refractive-index profile just like that of a GI multimode fiber [9,10]. The profile is expressed as

$$n(r) = n_0 \left(1 - \frac{1}{2} g^2 r^2 \right) \tag{11.7}$$

where g is the positive constant, n_0 is the refractive index on the axis, and r is the distance from the axis. The profile for the ideal lenslike medium for meridional rays is obtained with Fermat's principle and Snell's law. That is,

$$\int_s n(r) \, ds = \text{constant} \quad \text{(Fermat's principle)} \tag{11.8}$$

$$n(r_1) \cos \phi_1 = n(r_2) \cos \phi_2 = \cdots = n(r_m) \cos \phi_m \quad \text{(Snell's law)} \tag{11.9}$$

where the symbols ds, s, and ϕ are the infinitesimal path length, the ray path, and the angle, respectively, and they are indicated in Figure 11.5(a). The constant in (11.8) indicates that the integral is constant for all ray paths. The solution of (11.8) with (11.9) is

$$n(r) = n_0 \operatorname{sech}(gr) = n_0 \left\{ 1 + g^2 r^2 + \frac{2}{3} g^4 r^4 + \ldots \right\} \tag{11.10}$$

This index profile is the same with (11.7) when we ignore the higher order terms in the expansion. The rod lenses are fabricated by the ion exchange process. That is, the rod glass is immersed in a salt bath and the ion in a glass is exchanged by the ion in a salt bath. With careful control of the exchange time and the temperature, the desired graded refractive-index profile is obtained. The diameter of the rod lens is in the millimeter region.

The light propagating near the axis (the paraxial ray) in the medium expressed by (11.7) can be treated with geometric optics. It is convenient to express light propagation by using a ray matrix, such as

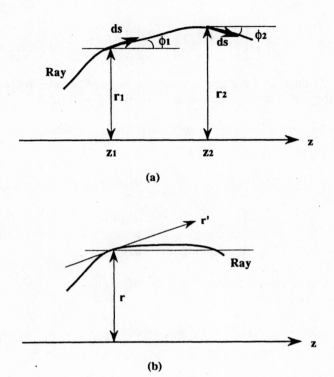

(a)

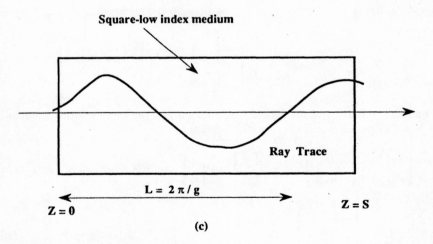

(b)

(c)

Figure 11.5 Ray propagation in a rod lens.

$$\begin{bmatrix} r \\ r' \end{bmatrix}_{out} = \begin{bmatrix} A & B \\ C & D \end{bmatrix} \begin{bmatrix} r \\ r' \end{bmatrix}_{in} \tag{11.11}$$

where r and r' are the distance from the axis and the slope of the ray, respectively. r' is expressed by the following equation in the case of Figure 11.5(b):

$$r' = \frac{dr}{dz} \tag{11.12}$$

The matrix is known as the ray matrix. In the case of a rod lens, it is

$$\begin{bmatrix} A & B \\ C & D \end{bmatrix} = \begin{bmatrix} \cos(gs) & \dfrac{1}{n_0 g} \sin(gs) \\ -n_0 g \sin(gs) & \cos(gs) \end{bmatrix} \tag{11.13}$$

The above ray matrix of a lenslike medium, which is like a rod lens, is derived using the following paraxial ray equation [11]:

$$\frac{\partial^2 r}{\partial z^2} = \frac{1}{n} \frac{\partial n}{\partial r} \tag{11.14}$$

The equation (11.13) indicates that the ray in a rod lens follows a sinusoidal path with period $L = 2\pi/g$, which is shown in Figure 11.5(c). For a quarter-period rod lens, which has a lens length of $s = L/4$ $(= \pi/2g)$, the ray matrix is expressed as

$$\begin{bmatrix} A & B \\ C & D \end{bmatrix} = \begin{bmatrix} 0 & \dfrac{1}{n_0 g} \\ -n_0 g & 0 \end{bmatrix} \tag{11.15}$$

The output lights for a quarter-period rod lens are expressed as follows when input light is a collimated beam or a focusing beam.

$$\begin{bmatrix} r \\ r' \end{bmatrix}_{out} = \begin{bmatrix} 0 \\ -n_0 g r \end{bmatrix} \quad \text{for} \quad \begin{bmatrix} r \\ r' \end{bmatrix}_{in} = \begin{bmatrix} r \\ 0 \end{bmatrix} \quad \text{(collimated input beam)} \tag{11.16}$$

$$\begin{bmatrix} r \\ r' \end{bmatrix}_{out} = \begin{bmatrix} \dfrac{1}{-n_0 g} r' \\ 0 \end{bmatrix} \quad \text{for} \quad \begin{bmatrix} r \\ r' \end{bmatrix}_{in} = \begin{bmatrix} 0 \\ r' \end{bmatrix} \quad \text{(focusing input beam)} \tag{11.17}$$

Therefore, the output beam is collimated for the focusing input beam, and the output beam is focusing for the collimated input beam.

11.1.2 Fiber-Type WDM Filter

There are several types of fiber-type WDM filters: fused-fiber couplers, polished fibers, fiber pieces using thin-film filters, and fibers with a interferometer configuration, and shown in Figures 11.6 and 11.7.

A fused-fiber coupler can be used as a fiber-type WDM filter when the coupling coefficient κ between two fibers in a fused-fiber coupler is designed for WDM. The coupling coefficient κ depends on the waveguide structure, such as the distance between two cores, the shape of core, and the refractive-index profile. By designing these waveguide parameters, fused-fiber couplers are used for WDM devices. The value of coupling coefficient κ for a WDM device is different from that of a 3-dB coupler. In the case of a 3-dB coupler, the value κ is nearly constant for a wavelength range. On the other hand, κ is variable for a wavelength range in the case of a WDM filter. The following equations for WDM couplers are the same as (10.29) and (10.30):

$$P_1(z) = |a_1(z)|^2 = A^2 \cos^2(\Delta\beta z) \tag{11.18}$$

$$P_2(z) = |a_2(z)|^2 = A^2 \sin^2(\Delta\beta z) \tag{11.19}$$

If $\kappa(\Delta\beta = |\kappa|)$ is designed with $z = L$, to hold that $\kappa L = 2\pi n + \pi/2$ for wavelength λ_1 and $\kappa L = 2\pi m + \pi$ for wavelength λ_2, where n and m are integers, then we obtain $P_1(L) = 0$ and $P_2(L) = A^2$ for λ_1, and $P_1(L) = A^2$ and $P_2(L) = 0$ for λ_2. This is the principle of WDM devices using a directional coupler, such as those in Figure 11.6(a,b). One example of a fused-fiber coupler for WDM use is reported in [12], and the reported isolation for a fabricated 1.32-/1.55-μm WDM coupler is 16 to 18 dB. The measured insertion loss is 0.05 dB and very small [12]. Generally, fused-fiber couplers, designed for both a wavelength-insensitive coupler and a WDM coupler, have very small insertion loss and stable properties against temperature change. A similar coupling mechanism takes place in the polished-fiber type of device as shown in Figure 11.5(b). By polishing the fiber side (not the fiber endface), two cores in two fibers become closer and coupling takes place between them. One example of couplers fabricated by polishing is reported in [13], and the reported isolation ranges from 50 to 10 dB, depending on the wavelength spacing (from 200 to 35 nm).

A fiber-type WDM device using a dielectric thin-film filter is possible when thin-film filters are inserted between fibers. One example of this type is shown in Figure 11.7(a). Thin-film filters are fabricated directly on the fiber endface by alternately evaporating TiO_2 and SiO_2 [14]. The reported insertion loss and isolation

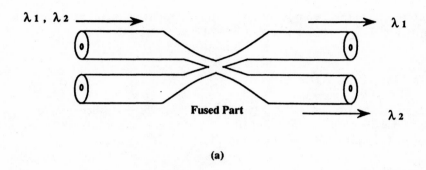

(a)

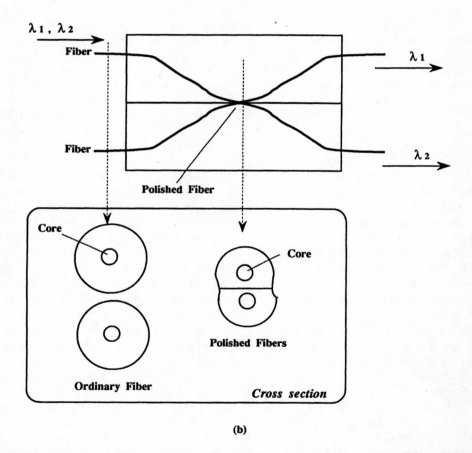

(b)

Figure 11.6 Fiber-type WDM device (1): (a) fused-coupler type; (b) fiber type using polished fibers.

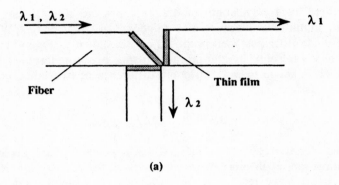

(a)

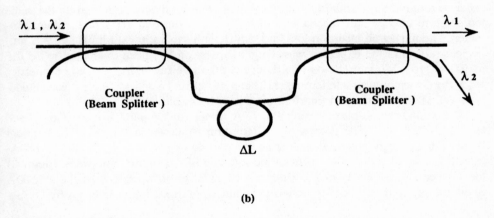

(b)

Figure 11.7 Fiber-type WDM device (2): (a) using thin-film filter; (b) using interferometer. (After [14,15].)

of a WDM device for a wavelength of 0.82/1.2 μm are about 1 and 40 dB, respectively, for a multimode fiber with a 60-μm core [14]. A fiber-type WDM device using a single-mode fiber has been proposed and is shown in Figure 11.7(b) [15]. It consists of a Mach-Zehnder interferometer using two couplers and fibers. The length ΔL, which is added to make a length difference of two arms, contributes a phase difference $\Delta \varphi$:

$$\Delta \varphi = 2\pi n_{\text{eff}} \Delta L / \lambda \qquad (11.20)$$

where n_{eff} is the effective refractive index, which depends on the wavelength. Two lightwaves with a phase difference are combined at the second coupler and they interfere. The interference results in the wavelength-dependent property used for wavelength multi/demultiplexers. It is reported that the output shows instability with time when constructed by two couplers with a 0.5 m-long fiber between them [15]. This is caused by the refractive-index change due to environmental changes of, for example, temperature and air current. A miniaturized configuration is adopted in order to remove the instability with time [15]. The miniaturization causes the same temperature change and mechanical disturbance in the two arms of the interferometer. The measured insertion loss and wavelength separation are 2 dB and 3.5 nm, respectively, and the expected wavelength separation has been evaluated to be 0.5 to 10 nm [15]. Miniaturization of a device is effective for realizing stable properties when the device uses an interferometer. Therefore, it can be anticipated that waveguide-type WDM filters using an interferometer will be stable.

A multi/demultiplexer using PANDA fibers for all parts has been proposed and fabricated [16]. The proposed configuration is shown in Figure 11.8. Incident lights with a linearly polarized state at an angle of 45 deg with respect to the PANDA-fiber principal axis are injected from the left side of the figure. The wavelengths of the incident lights are λ_1 to λ_N. Due to the modal birefringence B of the PANDA fiber, the phase difference $\Delta \theta$ between x- and y-polarized lights is given by

$$\Delta \theta = \frac{2\pi}{\lambda} BL \qquad (11.21)$$

This equation is derived by the following definition of birefringence B:

$$B \equiv \frac{\beta_x - \beta_y}{k} = \frac{\Delta \beta}{k} = \frac{\lambda \Delta \beta}{2\pi} \qquad (11.22)$$

where $\Delta \beta$ is the propagation constant difference between two polarized lightwaves with β_x and β_y. k is the wave number ($k = 2\pi / \lambda$). The equation (11.21) is derived by using (11.22) and $\Delta \theta = \Delta \beta L$. When the following condition expressed by (11.23)

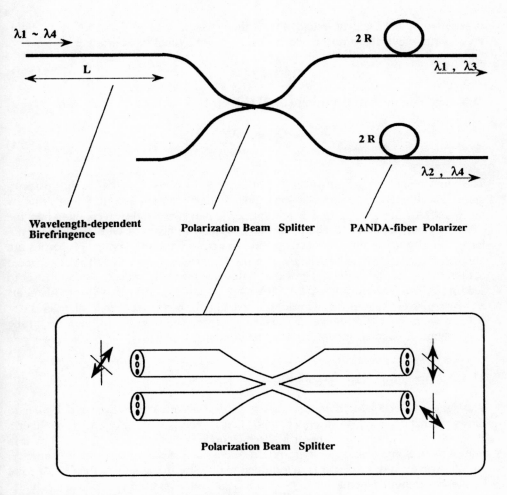

Figure 11.8 PANDA-fiber type of WDM device. (After [16].)

is satisfied by the incident N lightwaves, the principal axes of the neighboring lightwaves with wavelength difference $\Delta\lambda$ become orthogonal to each other.

$$\Delta\theta_{i+1} - \Delta\theta_i = \pi \tag{11.23}$$

This condition is rewritten using (11.21) and (11.23) as

$$L = \frac{\lambda^2}{2B\Delta\lambda} \tag{11.24}$$

The input neighboring lightwaves are orthogonal to each other based on this principle. The lightwaves with orthogonal polarization are split into different arms by the PANDA-fiber polarization beam splitter. Then these lightwaves enter into the PANDA-fiber polarizer to eliminate the unwanted lightwaves with orthogonal polarization. By using the polarizers, good separation (demultiplexing) properties are obtained. The value B of a PANDA fiber used in [16] is 5.1×10^{-4}. The experiment demonstrates the demultiplexing of two lightwaves with $\Delta\lambda = 1.38$ nm and L set to 1.65m. The extension ratio of PANDA-fiber polarizer is 36 dB with a coil radius $R = 4$ cm. Note that closer wavelength spacing $\Delta\lambda$ can be realized, such as $\Delta\lambda = 0.1$ nm when $L > 20$m; however, a stable performance may be difficult due to the temperature-dependent modal birefringent B [16].

11.1.3 Waveguide-Type WDM Filter

Waveguide-type WDM filters are also possible, and their principles are similar to those of fiber-type WDM filters [17–19]; that is, the basic configuration is either a simple directional coupler type or a Mach-Zehnder interferometer type as shown in Figure 11.9. Some waveguide-type WDM filters use a Mach-Zehnder interferometer, and the wavelength response is determined mainly by the length difference ΔL, not by the wavelength dependence of two directional couplers used in the Mach-Zehnder interferometer. The Mach-Zehnder interferometer is widely used for waveguide-type WDM filters, because a desirable κ value is easily fabricated with this interferometer design.

The principle of waveguide-type WDM filters using a simple directional coupler is the same as that explained in a fused-fiber coupler for WDM use in the previous section. The principle of waveguide-type WDM filters using a Mach-Zehnder interferometer is the same as that explained for the fiber-type filter shown in Figure 11.7(b). Here, some supplemental explanations are provided because of its importance. As shown in Figure 11.9, two lightwaves enter into the Mach-Zehnder interferometer, and we design these lightwaves to be separated at the two outputs, port 1 and port 2. Incident light is split by the first directional coupler and the phase

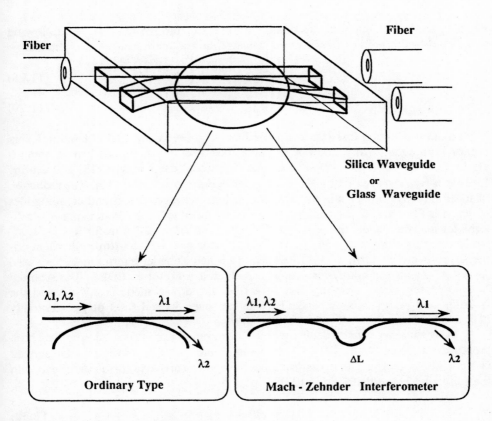

Figure 11.9 Waveguide-type WDM device.

difference of the split light is $\pi/2$. There is additional phase difference due to the length difference ΔL. By using (11.20), the total phase difference $\Delta\theta$ at the input of the second directional coupler is obtained as

$$\Delta\theta = \pi/2 + 2\pi n_{\text{eff}}\,\Delta L/\lambda \qquad (11.25)$$

We must design $\Delta\theta$ by taking into consideration the phase shift of $\pi/2$ by the second directional coupler, and the results are

$$\Delta\theta = 2n\pi + \pi/2 \quad (\text{for } \lambda_1) \qquad (11.26)$$

$$\Delta\theta = (2m + 1)\pi + \pi/2 \quad (\text{for } \lambda_2) \qquad (11.27)$$

where n and m are integers. By combining (11.25) through (11.27), the following equations for designing filters using a Mach-Zehnder interferometer are obtained:

$$n_{eff} \Delta L = n\lambda_1 \tag{11.28}$$

$$n_{eff} \Delta L = (m + 1/2)\lambda_2 \tag{11.29}$$

The reported insertion loss (fiber-to-fiber loss) of these WDM multi/demultiplexers ranges from about 3 to 5 dB. For example, insertion loss is reported to be about 5 dB for two waves in the case of a simple directional coupler type [18] and 2.6 dB for four waves in the case of a Mach-Zehnder interferometer type [19]. The reduction of insertion loss originates in the decrease of both waveguide loss and coupling loss between a fiber and a waveguide. As an example of ΔL for filters using a Mach-Zehnder interferometer, the reported ΔL values are $\Delta L = 2.7$ μm for a 1.3-/1.55-μm multi/demultiplexer, and $\Delta L = 15.5$ μm for a 1.5-/1.55-μm multi/demultiplexer, respectively [20]. WDM filters using a Mach-Zehnder interferometer are used for a 1.48-/1.55-μm multi/demultiplexer, which is used for an EDFA. The reported insertion loss, where coupling loss to a fiber is not included, is 0.7 dB and the isolation and polarization-dependent losses are over 30 and 0.03 dB, respectively [21].

Waveguide-type multi/demultiplexers based on an optical phased array has been proposed and demonstrated [22]. As shown in Figure 11.10(a), waveguides are formed in a curved geometry. The lights from the waveguides arrive at the output plane with the following phase distribution [23]:

$$\phi_i = \beta R_i\theta + \phi_0 \tag{11.30}$$

where ϕ_i and ϕ_0 are the phase of waveguide i and the initial phase, respectively. β, R_i, and θ are the propagation constant, the bending radius of waveguide i, and the angle of the concentric array, respectively. When we design the concentric waveguides with the following R_i, we can get the required phase distribution $\Phi(R_i)$ at the output plane, which results in the desired optical pattern.

$$\beta R_i\theta = \Phi(R_i) + 2\pi n_i \tag{11.31}$$

where n_i is an integer. If we select $\Phi(R_i) = 0$, the waveguide array behaves like a phased array with uniform phase distribution. The phased array shown in Figure 11.10(a) has dispersive characteristics, because the propagation constant β is different for different wavelength. This results in different distribution $\Phi(R_i)$ for different wavelengths, and the WDM filter is realized by using it. A four-channel demultiplexer in the 0.78-μm wavelength range with a channel spacing of 1.55 nm has been reported, and its configuration is shown in Figure 11.10(b) [22]. The ob-

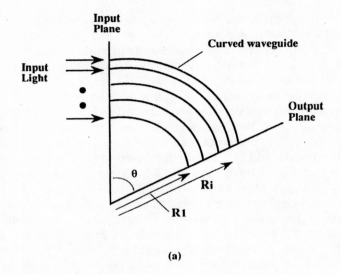

(a)

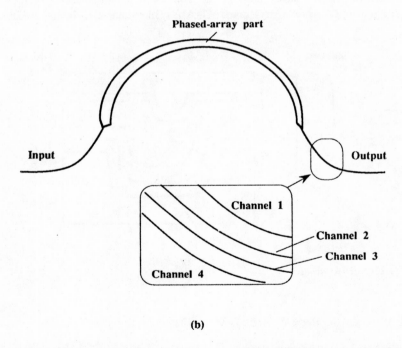

(b)

Figure 11.10 Waveguide-type WDM device based on an optical phased array: (a) phased-array configuration; (b) four-channel demultiplexer. (After [22,23].)

tained insertion losses excluding fiber coupling loss and the crosstalk value are 1.9 to 2.5 dB or 13 to 30 dB, depending on the channels and light polarization, respectively [22]. As in the case of a phased-array antenna in radio frequency (such as microwave frequency), several characteristic can be obtained in an optical frequency by designing the phase distribution $\Phi(W_i)$ at outputs of arrayed waveguides, where W_i indicates the ith individual waveguide in an array. Generally, a concentric array configuration is not necessary for obtaining the desired phase distribution $\Phi(W_i)$. In a radar application in radio frequency, a narrow beam is produced by a phased-array antenna. Similarly, an arrayed waveguide for a grating application has been proposed and demonstrated [24]. The configuration is shown in Figure 11.11. Any two adjacent waveguides have the same length difference ΔL, which results in the phase difference $2\pi n_{eff}\Delta L/\lambda$, where n_{eff} is the effective refractive index of the waveguide. The reported dispersion in a fabricated arrayed-waveguide grating in the 1.3-μm wavelength region is about 96 μm/nm at the x plane in Figure 11.11 [24]. In this demonstration, the number of waveguides in an array is 150, and the designed ΔL value is 17.54 μm. When fibers are aligned with a space of 250 μm between fibers at the x plane, the estimated wavelength spacing in a demultiplexer application is about 2.6 nm in this case. Waveguide-type multi/demultiplexers using semiconductors, such as GaAs/AlGaAs and InGaAlAs/InP have also been fabricated [25–27].

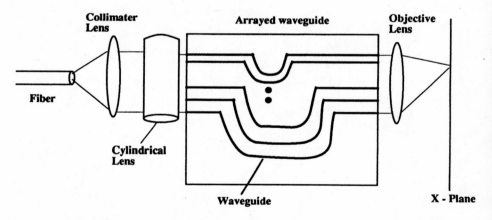

Figure 11.11 Arrayed-waveguide grating. (After [24].)

11.1.4 Waveguide-Type $N \times N$ WDM Filter

Waveguide-type $N \times N$ WDM filters, which are capable of multiplexing and demultiplexing a large number of input and output wavelengths simultaneously, have been proposed. The 7×7 and 11×11 WDM filters have been demonstrated using a planar SiO_2/Si waveguide [28,29], which consists of two star couplers and an arrayed-waveguide grating between the two star couplers. The configuration is shown in Figure 11.12 and the insertion loss, excluding fiber coupling losses, the crosstalk, and the wavelength spacing are reported to be lower than 2.5 dB, less than −25 dB, and 3.3 nm, respectively, for a 7 × 7 WDM filter [29]. A similar approach is used for a broadband WDM filter, and one demonstrated multiplexer has an insertion loss of about 2 dB, including fiber coupling loss, and crosstalk of about −35 dB for a two-wavelength application (λ = 1.3 and 1.55 μm) [30]. In this case, the 3-dB full bandwidths for λ = 1.3 and 1.55 μm are 78 and 93 nm, respectively.

Arrayed-waveguide Grating

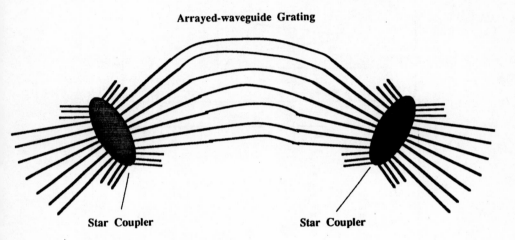

Star Coupler   Star Coupler

Figure 11.12 $N \times N$ WDM filter. (After [28,29].)

11.2 OFDM FILTER

It is common to use optical interference to realize an OFDM filter, which has much closer wavelength spacing than WDM filters do. As the result of the optical interference, a fine wavelength selectivity is realized. This selectivity appears periodically in the optical frequency (wavelength) domain. In this section, Fabry-Perot interferometer–type filters (FP filters), Mach-Zehnder interferometer–type filters (MZ filters) and ring resonator–type filters (RR filters) are explained as OFDM filters.

11.2.1 Fabry-Perot Interferometer–Type Filter

FP interferometers or FP resonators are used for several optical measurement systems and are used as laser diode resonators. A model of an FP interferometer is shown in Figure 11.13.

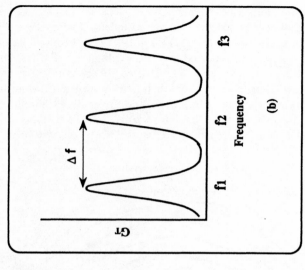

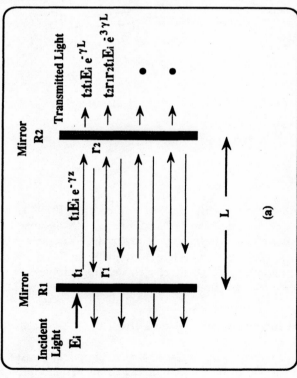

Figure 11.13 FP resonance model: (a) FP resonator; (b) transmittance profile.

Incident light enters the FP interferometer from the left side in Figure 11.13(a), and this light travels out from the right side. Some portion of light reflects at the right-side mirror, passes from the right side to the left side, and again reflects at the left-side mirror. Some portion of this double-path light reflects at the right-side mirror and passes, and so on. These multiple reflections are expressed as follows. The first transmitted light is $E_i t_1 t_2 e^{-\gamma L}$, the second is $E_i t_1 r_1 r_2 t_2 e^{-3\gamma L}$, and so on.

The notations E_i, t_1, t_2, r_1, r_2, γ, and L are the electric field of incident light, the transmittance coefficient at the left-side mirror, the transmittance coefficient at the right-side mirror, the reflection coefficient at the left-side mirror, the reflection coefficient at the right-side mirror, the complex propagation constant, and the FP length, respectively. The field of output light is the sum of these transmitted lights and is

$$E_t = E_i t_1 t_2 e^{-\gamma L}[1 + r_1 r_2 e^{-2\gamma L} + (r_1 r_2 e^{-2\gamma L})^2 + \ldots]$$

$$= \frac{t_1 t_2 e^{-\gamma L}}{1 - r_1 r_2 e^{-2\gamma L}} E_i \tag{11.32}$$

and the power transmittance G_T is

$$G_T = \frac{|E_t|^2}{|E_i|^2} = \frac{|t_1 t_2 e^{-\gamma L}|^2}{|1 - r_1 r_2 e^{-2\gamma L}|^2} \tag{11.33}$$

The propagation constant γ is

$$\gamma = \alpha/2 + j\beta \tag{11.34}$$

where α and β are the optical loss in the FP resonator and the phase constant, respectively. The phase constant β is $2\pi n/\lambda$, where n is the refractive index. Using (11.34), the calculated result of (11.33) is

$$G_T = \frac{(1 - R_1)(1 - R_2)G_s}{(1 - \sqrt{R_1 R_2}\, G_s)^2 + 4\sqrt{R_1 R_2}\, G_s \sin^2(\beta L)} \tag{11.35}$$

where the following relationships between reflection coefficients for amplitude and reflection coefficients for power are used:

$$R_1 = r_1^2 \qquad R_2 = r_2^2 \tag{11.36}$$

and G_s is defined as

$$G_s = \exp(-\alpha L) \tag{11.37}$$

No loss ($G_s = 1$) and $R_1 = R_2 = R$ are assumed, and then

$$G_T = \frac{(1 - R)^2}{(1 - R)^2 + 4R \sin^2(\beta L)} \tag{11.38}$$

We define Δf as the frequency interval between the maximum and minimum of the transmission power, so Δf is

$$\Delta f = c/(4n_{\text{eff}}L) \tag{11.39}$$

The bandwidth B (half width at half maximum (HWHM)) is

$$B = \frac{c}{2\pi n_{\text{eff}}L} \sin^{-1}\left(\frac{(1 - R)}{2\sqrt{R}}\right) \tag{11.40}$$

The ratio of Δf and B is commonly used as a measure of an FP filter, and the ratio is called finesse F. Finesse F is

$$F \equiv \frac{2\Delta f}{2B} = \frac{\pi}{2}\left[\sin^{-1}\left(\frac{1 - R}{2\sqrt{R}}\right)\right]^{-1} \tag{11.41}$$

and is approximately expressed for $R \sim 1$:

$$F \sim \frac{\pi\sqrt{R}}{1 - R} \tag{11.42}$$

For example, $F = 312$ for $R = 99\%$. To obtain a high-finesse filter, two-stage FP filters were considered for OFDM systems [31,32]. A tunable FP filter can be realized by adjusting the separation length L. There are several mechanisms for adjusting; one example is the use of a piezoelectric control device. Another way to obtain a tunable FP filter is the use of a liquid crystal, which is inserted into a cavity of an FP resonator [31]. With the use of a liquid crystal intracavity, a full width at half maximum (FWHM) of 0.17 to 0.35 nm and a wavelength-tunable range of over 50 nm at a 1.5-μm wavelength have been obtained [33]. A narrower bandwidth of HWHM < 10 GHz and a wavelength-tunable range of 9 nm from the use of a double-layer cavity structure were reported in [34].

11.2.2 Mach-Zehnder Interferometer–Type Filter

Although MZ filters can be made using several technologies, stable filters have been realized by using a planar waveguide technology. The configuration of an MZ filter is the same as the structure shown in Figure 10.8. The basic properties of MZ filters are understood by using the coupled-wave equation discussed in Chapter 10. As with the discussions of (10.26) and (10.27), the following equation holds for a coupler in an MZ filter:

$$
\begin{bmatrix} A_1(z) \\ A_2(z) \end{bmatrix} = \begin{bmatrix} \cos(\Delta\beta z) & j\sin(\Delta\beta z) \\ j\sin(\Delta\beta z) & \cos(\Delta\beta z) \end{bmatrix} \begin{bmatrix} A_1(0) \\ A_2(0) \end{bmatrix}
$$

$$
= \begin{bmatrix} t_1 & jr_1 \\ jr_1 & t_1 \end{bmatrix} \begin{bmatrix} A_1(0) \\ A_2(0) \end{bmatrix} \tag{11.43}
$$

where A_1 and A_2 are the amplitudes defined in (10.16) and (10.17), and $t_1 = \cos(\Delta\beta z)$ and $r_1 = \sin(\Delta\beta z)$ are introduced. Here we assume that the coupler is lossless and $\beta_1 = \beta_2 = \beta$. The phase shift part of an MZ filter has a length difference ΔL between two arms, and it is expressed by

$$
\begin{bmatrix} \exp(-j\beta\Delta L/2) & 0 \\ 0 & \exp(j\beta\Delta L/2) \end{bmatrix} \tag{11.44}
$$

The following equation is obtained for an MZ filter, which has two couplers and one phase shifter between them:

$$
\begin{bmatrix} A_1(z) \\ A_2(z) \end{bmatrix} = [M] \begin{bmatrix} A_1(0) \\ A_2(0) \end{bmatrix} \tag{11.45}
$$

$$
[M] = \begin{bmatrix} t_2 & jr_2 \\ jr_2 & t_2 \end{bmatrix} \begin{bmatrix} \exp(-j\beta\Delta L/2) & 0 \\ 0 & \exp(j\beta\Delta L/2) \end{bmatrix} \begin{bmatrix} t_1 & jr_1 \\ jr_1 & t_1 \end{bmatrix} \tag{11.46}
$$

In only one input case ($A_2(0) = 0$), the following equations are obtained by using the equations above.

$$
\frac{P_{1\,out}}{P_{1\,in}} = \left| \frac{A_1(z)}{A_1(0)} \right|^2 = T_1 T_2 + R_1 R_2 - 2\sqrt{T_1 T_2 R_1 R_2}\, \cos(\beta\Delta L)
$$

$$
= (\sqrt{R_1 R_2} - \sqrt{T_1 T_2})^2 + 4\sqrt{T_1 T_2 R_1 R_2}\, \sin^2(\beta\Delta L/2) \tag{11.47}
$$

$$\frac{P_{2\text{out}}}{P_{1\text{in}}} = \left|\frac{A_2(z)}{A_1(0)}\right|^2 = T_1R_2 + R_1T_2 + 2\sqrt{T_1T_2R_1R_2}\,\cos(\beta\Delta L)$$

$$= (\sqrt{T_1R_2} - \sqrt{R_1T_2})^2 + 4\sqrt{T_1T_2R_1R_2}\,\cos^2(\beta\Delta L/2) \tag{11.48}$$

where

$$R_1 = r_1^2, \qquad R_2 = r_2^2 \tag{11.49}$$

$$T_1 = 1 - R_1, \qquad T_2 = 1 - R_2 \tag{11.50}$$

The frequency space Δf between the maximum transmission power and the minimum transmission power of an MZ filter is derived by using (11.47) and (11.48). That is,

$$\cos(\beta\Delta L) = 1 \quad \text{and} \quad 0 \tag{11.51}$$

Therefore,

$$\beta\Delta L = 2\pi N \quad \text{and} \quad 2\pi N + \pi \tag{11.52}$$

holds, where N is an integer. With (11.52) and (11.53), Δf is obtained.

$$\beta = \frac{2\pi n_{\text{eff}}f}{c} \tag{11.53}$$

$$\Delta f = f_2 - f_1 = \frac{c}{2\pi n_{\text{eff}}\Delta L}(2\pi N + \pi - 2\pi N) = \frac{c}{2n_{\text{eff}}\Delta L} \tag{11.54}$$

where n_{eff} is the effective refractive index of a waveguide and c is the light velocity in a vacuum. Extinction ratios E_1 and E_2 of two outputs are derived by using (11.47) and (11.48) as

$$E_1 = -10\log[(\sqrt{R_1R_2} - \sqrt{T_1T_2})^2/(\sqrt{R_1R_2} + \sqrt{T_1T_2})^2] \quad \text{(dB)} \tag{11.55}$$

$$E_2 = -10\log[(\sqrt{R_1T_2} - \sqrt{R_2T_1})^2/(\sqrt{R_1T_2} + \sqrt{R_2T_1})^2] \quad \text{(dB)} \tag{11.56}$$

In the case of $T_1 = T_2 = R_1 = R_2 = 0.5$ (3-dB coupler), the extinction ratios $E_1 = E_2 = \infty$. A four-channel filter with 5-GHz spacing in the 1.5-μm wavelength region has been reported, constructed by using a two-stage MZ filter fabricated by a silica-based planar waveguide technology [35]. To obtain $\Delta f = 5$ GHz ($\Delta f = 5$ GHz corresponds to $\Delta\lambda = 0.04$ nm in 1.5-μm wavelength region), ΔL is set to 2.04 cm. The periodic property of an MZ filter has been confirmed experimentally for 14 nm (about 1,750 GHz) in the 1.5-μm wavelength range [36]. A tunable filter is realized by forming a thin-film heater in one arm of a Mach-Zehnder interferometer. With two-stage MZ filters, a 5-GHz-spaced, eight-channel tunable filter with a tuning speed of 0.8 ms was reported [37].

Besides an MZ filter, Mach-Zehnder interferometer–type devices such as a LiNbO$_3$ modulator have been widely used for optical switching and modulation applications. The principle of their use is based on phase change at the phase shifter of a Mach-Zehnder arm by electrical means.

11.2.3 Ring Resonator–Type Filter

A ring resonator has been used for channel-dropping or branching filters at microwave frequencies. An optical filter using a ring resonator was discussed at the early investigation stage [38]. Several types of ring resonators have been proposed and fabricated [39–43]. The configuration of filters using single- and double-ring resonators are shown in Figure 11.14(a,b).

Here we treat a single resonator of Figure 11.14(a). The input lightwave in a straight waveguide couples to a ring waveguide, and the coupled lightwave travels around a ring. The coupling is modeled by a directional coupler, where amplitudes are treated in a coupled-wave equation [40]. That is,

$$\begin{bmatrix} A_3 \\ A_4 \end{bmatrix} = (1 - \Gamma)^{1/2} \begin{bmatrix} \cos(\Delta\beta Z_0) & j\sin(\Delta\beta Z_0) \\ j\sin(\Delta\beta Z_0) & \cos(\Delta\beta Z_0) \end{bmatrix} \begin{bmatrix} A_1 \\ A_2 \end{bmatrix} \tag{11.57}$$

where Γ and Z_0 are the intensity loss and the coupling length, respectively. The port numbers are indicated in Figure 11.14(a). The coupled lightwave suffers a phase shift βL_r and an attenuation α by single-pass travel, where L_r is the ring pass length ($L_r = 2\pi r_r$: r_r is the radius of a ring). This is expressed as

$$A_2 = A_4 e^{-\alpha L_r} e^{-j\beta L_r} \tag{11.58}$$

With (11.57) and (11.58), the power ratio of P_3/P_1 is obtained.

$$\frac{P_3}{P_1} = \left| \frac{A_3}{A_1} \right|^2 = \left| (1 - \Gamma)^{1/2} \frac{\cos(\Delta\beta Z_0) - (1 - \Gamma)^{1/2} e^{-\alpha L_r} e^{-j\beta L_r}}{1 - (1 - \Gamma)^{1/2} \cos(\Delta\beta Z_0) e^{-\alpha L_r} e^{-j\beta L_r}} \right|^2$$

$$\tag{11.59}$$

$$= (1 - \Gamma) \left[1 - \frac{[1 - (1 - \Gamma)e^{-2\alpha L_r}][1 - \cos^2(\Delta\beta Z_0)]}{[1 - (1 - \Gamma)^{1/2} e^{-\alpha L_r} \cos(\Delta\beta Z_0)]^2 + 4(1 - \Gamma)^{1/2} \cos(\Delta\beta Z_0) \sin^2(\beta L_r/2)} \right]$$

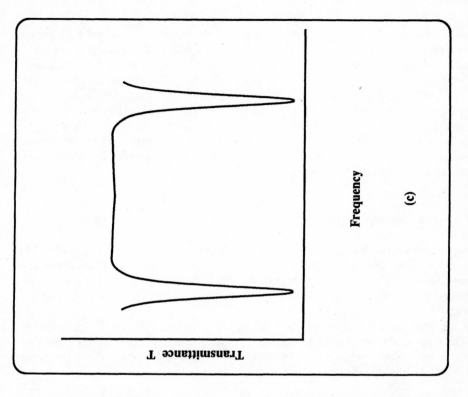

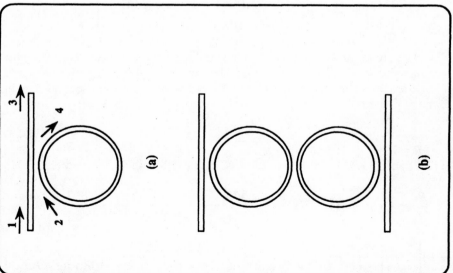

Figure 11.14 Optical filter using ring resonator: (a) single-ring resonator; (b) double-ring resonator; (c) transmittance profile.

The constructive and destructive interferences in a ring occur under the following conditions:

$$\beta L_r/2 = \pi N \quad \text{and} \quad \pi N + \pi/2 \tag{11.60}$$

where N is an integer. Then the frequency space Δf between the maximum and minimum of the transmission power is obtained as

$$\Delta f = \frac{c}{2n_{\text{eff}}L_r} \tag{11.61}$$

The MZ filters have square sine- or cosine-shaped transmittance expressed in (11.47) and (11.48). Therefore, they have insufficient wide passband and sharp cutoff characteristics for some applications. The RR filters have the potential to have wider passband and sharper cutoff characteristics compared to MZ filters. RR filters using double-ring configurations with a free spectral range (FSR) of 37.2 GHz and a finesse of higher than 182 were reported in [43], and they were fabricated by a silica-based planar waveguide technology.

11.3 BAND-REJECTION FILTERS AND BRAGG REFLECTORS

BRFs can be constructed using several technologies, such as multilayer dielectric thin films. Generally, a BPF can be considered as a BRF for the rejected lightwave in a BPF. These types of BRFs were already discussed in the previous sections. In this section, we discuss only a Bragg reflector as an application to a BRF because of its sharp characteristics.

A Bragg reflector is formed in a fiber or in a planar waveguide, and strong wavelength-selective reflection takes place [44,45]. A Bragg reflector is used in a laser diode, and the laser diode using this reflector is known as a DFB LD and a distributed Bragg reflector (DBR) laser diode. Derivation of the reflection coefficient in the Bragg reflector is based on the coupled-mode equation, which is derived in Chapter 10.

$$\frac{dA_1(z)}{dz} = j\kappa_1 A_2(z)e^{j(\beta_1 - \beta_2)z} \tag{11.62}$$

$$\frac{dA_2(z)}{dz} = j\kappa_2 A_1(z)e^{-j(\beta_1 - \beta_2)z} \tag{11.63}$$

where β_1 and β_2 are the phase constants (time dependence of $\exp[j(\omega t - \beta z)]$ is assumed) for waveguides WG_1 and WG_2, respectively. In the case of the Bragg reflector modeled in Figure 11.15, WG_1 and WG_2 are the same waveguide; that is, we treat the coupling of two modes with β_1 and β_2 in the same waveguide. The two modes propagate in opposite directions because of a reflector. As the coupling of counterpropagation modes, $\beta_1\beta_2 < 0$ holds. The refractive index varies in the Bragg reflector with period Λ in the model. When $f(z)$ represents the refractive index, the following equation holds:

$$f(z) = f(z_0 + m\Lambda) \tag{11.64}$$

where m is an integer. Since $f(z)$ is the periodic function, it is expressed by the following Fourier expansion in space with the spatial frequency $1/\Lambda$ with a constant C_S.

$$f(z) = \sum_S C_S e^{j2\pi Sz/\Lambda} \tag{11.65}$$

As the refractive index changes, the coupling coefficient κ changes in a similar manner. For the case of $S = 1$, the coupling coefficient κ_1 is

$$\kappa_1 = \kappa_B e^{-j2\pi z/\Lambda} \tag{11.66}$$

where κ_B is the real constant. In this case, the coupled-mode equations for a lossless waveguide are

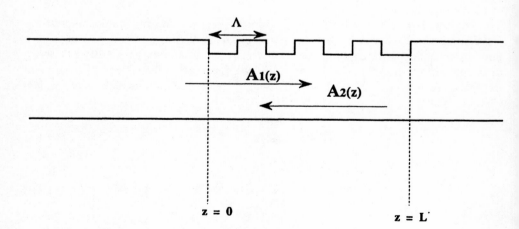

Figure 11.15 Model of the Bragg reflector.

$$\frac{dA_1(z)}{dz} = j\kappa_B A_2(z) e^{j(\beta_1 - \beta_2 - 2\pi/\Lambda)z} \tag{11.67}$$

$$\frac{dA_2(z)}{dz} = -j\kappa_B A_1(z) e^{-j(\beta_1 - \beta_2 - 2\pi/\Lambda)z} \tag{11.68}$$

As two waves propagate in a counterdirection, the condition of a lossless waveguide is

$$\frac{d}{dz}[|A_1(z)|^2 - |A_2(z)|^2] = 0 \tag{11.69}$$

and this results in $\kappa_1 = -\kappa_2^*$, which is used for deriving (11.67) and (11.68). When we consider the same mode with counterpropagation, the following equation holds.

$$\beta_1 = -\beta_2 = \beta = n_{eff}(2\pi/\lambda) \tag{11.70}$$

where n_{eff} is the effective refractive index. When the following condition holds, strong reflection occurs.

$$\beta = \pi/\Lambda \tag{11.71}$$

In this case, (11.67) and (11.68) are

$$\frac{dA_1(z)}{dz} = j\kappa_B A_2(z) \tag{11.72}$$

$$\frac{dA_2(z)}{dz} = -j\kappa_B A_1(z) \tag{11.73}$$

With the boundary condition of $A_1(0) = A$ and $A_2(L) = 0$, the solutions are

$$R \equiv \left|\frac{A_2(0)}{A_1(0)}\right|^2 = \tanh^2(\kappa_B L) \tag{11.74}$$

$$T \equiv \left|\frac{A_1(L)}{A_1(0)}\right|^2 = \text{sech}^2(\kappa_B L) \tag{11.75}$$

where R and T are the reflection and the transmission coefficients, respectively. The condition (11.71) is rewritten as

$$n_{\text{eff}}(2\pi/\lambda) = \pi/\Lambda \qquad (11.76)$$

and this results in

$$\lambda_B = 2n_{\text{eff}}\Lambda \qquad (11.77)$$

The band-rejection filter or reflector is realized when the above condition is met, with which strong reflection occurs. For example, $\kappa_B L = 3$ is obtained for $R = 0.99$ by using (11.74).

Several Bragg filters formed in a fiber have been discussed, and some of them are reported in [46,47]. A diffraction grating is formed in a polished cladding region by etching in these cases. It has been reported that the fabricated reflector has 25% reflection efficiency with 0.6-nm bandwidth (FWHM) in the 1.5-μm wavelength region [47].

An interesting method for forming a Bragg reflector in a fiber was developed in 1978 [48]. In this method, a fiber is exposed in a laser and light-induced refractive-index changes take place in a core. The reported experiments use a 1m germanium-doped fiber with a core diameter of 2.5 μm as a fiber, and an argon laser with 250 mW of a 0.488-μm wavelength as a laser. The reflection efficiency is reported to be 44%. It is considered that the Bragg reflectors are formed in a fiber by photo-refractivity [48,49]. A holographic method for forming Bragg reflectors in a fiber has been reported. The fabricated filters have a reflection efficiency of 50% to 55% and a bandwidth of 42 GHz (FWHM) at a 0.58-μm wavelength [50]. A grating operating at the 1.5-μm wavelength region has also been reported, and the fabricated filters have a reflection efficiency of 10% and a bandwidth of about 1 nm [51]. An EDF laser with a 1.5-μm wavelength has been demonstrated using the fabricated grating [51].

11.4 OPTICAL SIGNAL-PROCESSING FILTER

The WDM filters, OFDM filters, and BRFs explained in the previous sections are analogous to LPFs, high-pass filters (HPF), BPFs, and BRFs in the electric circuit and microwave fields. These optical filters are used for selecting a desired lightwave among many lightwaves, including noisy background lightwaves. Many electric filters are used for selecting a desired electric wave in a microwave region. Besides these filters, several other filters have been used for signal processing in electric circuit and microwave fields. Electric filters for equalizing and noise reduction are commonly used for metallic-cable or microwave transmission systems. They are also used for fiber-optic transmission systems after being converted from optical signals to electrical signals. Optical filters used for signal processing in the optical domain (without being converted to electric signals) have been investigated. An optical fiber or optical waveguide can be considered as a delay line, and a directional coupler and a star coupler can be considered as a branching (tapping) or summing (combining) element. Since the refractive index is about 1.46 for a fiber, the delay is about 5 ns/m. The use of these elements has allowed transversal filters and several lattice structure filters to be proposed and discussed [52–57].

The basic configuration of a transversal filter is shown in Figure 11.16(a). This filter is a tapped-delay-line form and the delay time is τ. The taps are weighted by the coefficient of C_i in this figure. An optical transversal filter has been realized by using a fiber in a V-groove at the early investigating stage [54]. For simplicity and stable operation, an optical transversal filter using a silica-based waveguide has been developed [57]. The fabricated transversal filter with four taps is shown in Figure 11.16(b) [57]. The tap weighting coefficient is expressed by a complex coefficient

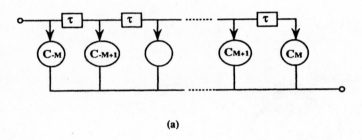

(a)

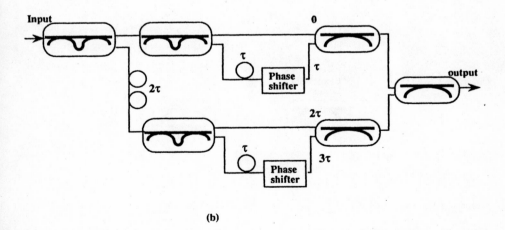

(b)

Figure 11.16 Transversal filter: (a) basic configuration; (b) fabricated optical filter (silica-based waveguide). (After [57].)

C_i, and this is realized by the splitter and the phase shifter. In this case, the splitter is made by a Mach-Zehnder type, and the phase shifter is composed of a waveguide with a thin-film heater attached to it. The phase is shifted by the thermo-optic effect. The unit delay line is a waveguide with a length of 4 cm ($\tau = 200$ ps). An HPF was fabricated and a differentiation of optical signals (50 Mbps) has been demonstrated by using this HPF [57].

Optical signal processing using a lattice structure filter has been proposed and the basic elements for the lattice filter are shown in Figure 11.17 [56]. One such filter is a finite-duration impulse response (FIR) filter, a nonrecursive system in which present output is a function of past and present inputs. Another type of filter is an infinite-duration impulse response (IIR) filter, a recursive system in which the present output is a function of past outputs, as well as past and present inputs. These filters are optically constructed by the elements shown in Figure 11.17.

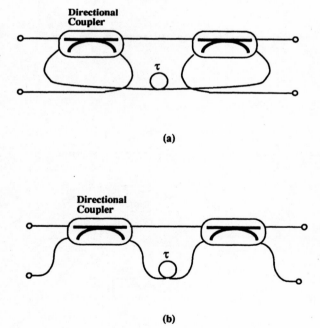

Directional
Coupler

(a)

Directional
Coupler

(b)

Figure 11.17 Basic elements for fiber-optic lattice: (a) recursive element; (b) nonrecursive element. (After [56].)

REFERENCES

[1] Born, M., and E. Wolf, Principles of Optics, Pergamon Press, 1974.

[2] Collin, R. E., Foundations of Microwave Engineering," MacGraw-Hill, 1966.

[3] Minowa, J., and Y. Fujii, "Dielectric Multilayer Thin-Film Filters for WDM Transmission Systems," IEEE J. Lightwave Technol., Vol. LT-1, 1983, p. 116.

[4] Watanabe, R., Y. Fujii, K. Nosu, and J. Minowa, "Optical Multi/Demultiplexers for Single-Mode Fiber Transmission," IEEE J. Quantum Electron., Vol. QE-17, 1981, p. 974.

[5] Seki, M., K. Kobayashi, Y. Odagiri, M. Shikada, T. Tanigawa, and R. Ishikawa, "20-Channel Micro-Optic Grating Demultiplexer for 1.1–1.6 μm Band Using a Small Focusing Parameter Graded-Index Rod Lens," Electron. Lett., Vol. 18, 1982, p. 257.

[6] Nosu, K., H. Ishio, and K. Hashimoto, "Multireflection Optical Multi/Demultiplexer Using Interference Filters," Electron. Lett., Vol. 15, 1979, p. 414.

[7] Kawachi, M., Y. Yamada, M. Yasu, and M. Kobayashi, "Guided-Wave Optical Wavelength-Division Multi/Demultiplexer Using High-Silica Channel Waveguides," Electron. Lett., Vol. 21, 1985, p. 314.

[8] Seki, M., R. Sugawara, Y. Handa, E. Okuda, H. Wada, and T. Yamasaki, "High-Performance Guided-Wave Multi/Demultiplexer Based on Novel Design Using Embedded Gradient-Index Waveguides in Glass," Electron. Lett., Vol. 23, 1987, p. 948.

[9] Uchida, T., M. Furukawa, I. Kitano, K. Koizumi, and H. Matsumura, "A Light-Focusing Fiber Guide," IEEE J. Quantum Electron., Vol. QE-5, 1969, p. 331.

[10] Uchida, T., M. Furukawa, I. Kitano, K. Koizumi, and H. Matsumura, "Optical Characteristics of a Light-Focusing Fiber Guide and Its Applications," IEEE J. Quantum Electron., Vol. QE-6, 1970, p. 606.

[11] Miller, S. E., "Light Propagation in Generalized Lens-Like Media," Bell Syst., Tech. J., Vol. 44, 1965, p. 2017.

[12] Lawson, C. M., P. M. Kopera, T. Y. Hsu, and V. J. Tekippe, "In-Line Single-Mode Wavelength Division Multiplexer/Demultiplexer," Electron. Lett., Vol. 20, 1984, p. 963.

[13] Digonnet, M., and H. J. Shaw, "Wavelength Multiplexing in Single-Mode Fiber Couplers," Appl. Opt., Vol. 22, 1983, p. 484.

[14] Miyauchi, E., T. Iwama, H. Nakajima, N. Tokoyo, and K. Terai, "Compact Wavelength Multiplexer Using Optical-Fiber Pieces," Opt. Lett., Vol. 5, 1980, p. 321.

[15] Sheem, S. K., and R. P. Moeller, "Single-Mode Fiber Wavelength Multiplexer," J. Appl. Phys., Vol. 51, 1980, p. 4050.

[16] Okamoto, K., T. Morioka, I. Yokohama, and J. Noda, "All-PANDA-Fiber Multi/Demultiplexer Utilizing Polarization Beat Phenomenon in Birefringent Fibers," Electron. Lett., Vol. 22, 1986, p. 181.

[17] Takato, N., Y. Otsuka, K. Jinguji, M. Yasu, and M. Kawachi, "Low-Loss Directional Coupler Using High-Silica Embedded Channel Waveguides," Optoelectronics Conf. (OEC'86), paper A3–3, 1986.

[18] Imoto, K., H. Sano, and M. Miyazaki, "Guided-Wave Multi/Demultiplexers With High Stopband Rejection," Appl. Opt., Vol. 26, 1987, p. 4214.

[19] Verbeek, B. H., C. H. Henry, N. A. Olsson, K. J. Orlowsky, R. F. Kazarinov, and B. H. Johnson, "Integrated Four-Channel Mach-Zehnder Multi/Demultiplexer Fabricated With Phosphorous Doped SiO$_2$ Waveguides on Si," IEEE J. Lightwave Technol., Vol. 6, 1988, p. 1011.

[20] Kominato, T., M. Yasu, N. Takato, and M. Kawachi, "Optical WDM Circuits With Guided-Wave Mach-Zehnder Interferometer Configuration," IEICE of Japan, Natl. Conf. Rec. on Commun., No. C-502, Spring 1989 (in Japanese).

[21] Arai, H., H. Uetsuka, H. Okano, H. Hakuta, and K. Akiba, "Fabrication of Planar Lightwave Circuit 1.48 μm/1.55 μm WDM for EDFA," IEICE of Japan, Natl. Conf. Rec. on Commun., No. C-213, Autumn 1993 (in Japanese).

[22] Vellekoop, A. R., and M. K. Smit, "Four-Channel Integrated-Optic Wavelength Demultiplexer With Weak Polarization Dependence," IEEE J. Lightwave Technol., Vol. 9, 1991, p. 310.

[23] Smit, M. K., "New Focusing and Dispersive Planar Component Based on an Optical Phased Array," Electron. Lett., Vol. 24, 1988, p. 385.

[24] Takahashi, H., S. Suzuki, K. Kato, and I. Nishi, "Arrayed-Waveguide Grating for Wavelength Division Multi/Demultiplexer With Nanometer Resolution," Electron. Lett., Vol. 26, 1990, p. 87.

[25] Deri, R. J., A. Yi-Yan, R. J. Hawkins, and M. Seto, "GaAs/AlGaAs Integrated-Optic Wavelength Demultiplexer," Opt. Lett., Vol. 13, 1988, p. 1047.

[26] Croston, I. R., and T. P. Young, "Design of an InGaAlAs/InP '3mi' Wavelength Division Demultiplexer Employing a Novel Mode Transformer," Electron. Lett., Vol. 26, 1990, p. 336.

[27] Wakatsuki, A., K. Okamoto, and O. Mikami, "Proposed Semiconductor Narrow-Band Wavelength Filter Using Directional Coupler," Electron. Lett., Vol. 26, 1990, p. 1573.

[28] Dragone, C., "An $N \times N$ Optical Multiplexer Using a Planar Arrangement of Two Star Couplers," Photon. Technol. Lett., Vol. 3, 1991, p. 812.

[29] Dragone, C., C. A. Edwards, and R. C. Kistler, "Integrated Optics $N \times N$ Multiplexer on Silicon," Photon. Technol. Lett., Vol. 3, 1991, p. 896.

[30] Adar, R., C. H. Henry, C. Dragone, R. C. Kistler, and M. A. Milbrodt, "Broad-Band Array Multiplexers Made With Silica Waveguides on Silicon," IEEE J. Lightwave Technol., Vol. 11, 1993, p. 212.

[31] Mallinson, S. R., "Wavelength-Selective Filters for Single-Mode Fiber WDM Systems Using Fabry-Perot Interferometers," Appl. Opt., Vol. 26, 1987, p. 430.

[32] Saleh, A. A. M., and J. Stone, "Two-Stage Fabry-Perot Filters as Demultiplexers in Optical FDMA LAN's," IEEE J. Lightwave Technol., Vol. 7, 1989, p. 323.

[33] Hirabayashi, K., H. Tuda, and T. Kurokawa, "Narrow-Band Tunable Wavelength-Selective Filters of Fabry-Perot Interferometers With a Liquid Crystal Intracavity," Photon. Technol. Lett., Vol. 3, 1991, p. 213.

[34] Hirabayashi, K., H. Tuda, and T. Kurokawa, "New Structure of Tunable Wavelength-Selective Filters With a Liquid Crystal for FDM Systems," Photon. Technol. Lett., Vol. 3, 1991, p. 741.

[35] Inoue, K., N. Takato, H. Toba, and M. Kawachi, "A Four-Channel Optical Waveguide Multi/Demultiplexer for 5-GHz Spaced Optical FDM Transmission," IEEE J. Lightwave Technol., Vol. 6, 1988, p. 339.

[36] Kashima, N., "Upgrade of Passive Optical Subscriber Network," IEEE J. Lightwave Technol., Vol. 9, 1991, p. 113.

[37] Toba, H., K. Oda, N. Takato, and K. Nosu, "5-GHz Spaced, Eight-Channel, Guided-Wave Tunable Multi/Demultiplexer for Optical FDM Transmission Systems," Electron. Lett., Vol. 23, 1987, p. 788.

[38] Marcatili, E. A. J., "Bends in Optical Dielectric Guides," Bell Syst. Tech. J., Vol. 48, 1969, p. 2103.

[39] Haavisto, J., and G. A. Pajer, "Resonance Effects in Low-Loss Ring Waveguides," Opt. Lett., Vol. 5, 1980, p. 510.

[40] Stokes, L. F., M. Chodorow, and H. J. Shaw, "All-Single-Mode Fiber Resonator," Opt. Lett., Vol. 7, 1982, p. 288.

[41] Walker, R. G., and C. D. W. Wilkinson, "Integrated Optical Ring Resonators Made by Silver Ion-Exchange in Glass," Appl. Opt., Vol. 22, 1983, p. 1029.

[42] Oda, K., N. Takato, H. Toba, and K. Nosu, "A Wide-Band Guided-Wave Periodic Multi/Demultiplexer With a Ring Resonator for Optical FDM Transmission Systems," IEEE J. Lightwave Technol., Vol. 6, 1988, p. 1016.

[43] Oda, K., N. Takato, and H. Toba, "A Wide-FSR Waveguide Double-Ring Resonator for Optical FDM Transmission Systems," IEEE J. Lightwave Technol., Vol. 9, 1991, p. 728.

[44] Yariv, A., Quantum Electronics, John Wiley & Sons, 1975.

[45] Yariv, A., and M. Nakamura, "Periodic Structures for Integrated Optics," IEEE J. Quantum Electron., Vol. QE-13, 1977, p. 233.

[46] Russel, P. St. J., and R. Ulrich, "Grating-Fiber Coupler as a High-Resolution Spectrometer," Opt. Lett., Vol. 10, 1985, p. 291.

[47] Whalen, W. S., M. D. Divino, and R. C. Alferness, "Demonstration of a Narrowband Bragg-Reflection Filter in a Single-Mode Fiber Directional Coupler," Electron. Lett., Vol. 22, 1986, p. 681.

[48] Hill, K. O., Y. Fujii, D. C. Johnson, and B. S. Kawasaki, "Photosensitivity in Optical Fiber Waveguides: Application to Reflection Filter Fabrication," Appl. Phys. Lett., Vol. 32, 1978, p. 647.

[49] Stone, J., "Photorefractivity in GeO_2-Doped Silica Fibers," J. Appl. Phys., Vol. 62, 1987, p. 4371.

[50] Meltz, G., W. W. Morey, and W. H. Glenn, "Formation of Bragg Gratings in Optical Fibers by a Transverse Holographic Method," Opt. Lett., Vol. 14, 1989, p. 823.

[51] Kashyap, R., J. R. Armitage, R. Wyatt, S. T. Davey, and D. L. Williams, "All-Fiber Narrowband Reflection Gratings at 1500 nm," Electron. Lett., Vol. 24, 1990, p. 730.

[52] Wilner, K., and A. P. Van Den Heuvel, "Fiber-Optic Delay Lines for Microwave Signal Processing," Proc. IEEE, Vol. 64, 1976, p. 805.

[53] Tur, M., J. W. Goodman, B. Moslehi, J. E. Bowers, and H. J. Shaw, "Fiber-Optic Signal Processor With Application to Matrix-Vector Multiplication and Lattice Filtering," Opt. Lett., Vol. 7, 1982, p. 463.

[54] Newton, S. A., K. P. Jackson, and H. J. Shaw, "Optical Fiber V-Groove Transversal Filter," Appl. Phys. Lett., Vol. 43, 1983, p. 149.

[55] Moslehi, B., J. W. Goodman, M. Tur, and H. J. Shaw, "Fiber-Optic Lattice Signal Processing," Proc. IEEE, Vol. 72, 1984, p. 909.

[56] Jackson, K. P., S. A. Newton, B. Moslehi, M. Tur, C. C. Cutler, J. W. Goodman, and H. J. Shaw, "Optical Fiber Delay-Line Signal Processing," IEEE Trans. Microwave Theory and Technol., Vol. MTT-33, 1985, p. 193.

[57] Sasayama, K., M. Okuno, K. Habara, "Coherent Optical Transversal Filter Using Silica-Based Single-Mode Waveguides," Electron. Lett., Vol. 25, 1989, p. 1508.

Chapter 12
Attenuators, Isolators, and Circulators

Optical attenuators, isolators, and circulators are used for improving the system performance just like the analogous components used in a radio or microwave frequency field. The principles used in these components are similar to those used in a radio or microwave frequency field. Most passive optical components are reciprocal; however, both the isolator and the circulator are nonreciprocal components. Also explained in this chapter are fiber-type polarizers, which have low insertion loss compared to the ordinary bulk-type polarizers.

12.1 OPTICAL ATTENUATORS

Optical attenuators decrease light intensity. They are used in several applications, such as fiber-optic transmission systems, in which they prevent the saturation of a receiver in the case of a strong intensity light. Several types of attenuators are shown in Figure 12.1. Many of them use materials to absorb light (Figs. 12.1(a,b)). Since the absorption generally depends on the wavelength, the value of attenuation also depends on the wavelength. For decreasing reflection intensity, a thin-film coating (multilayered dielectric thin-film coating) is used at the surface of the absorption materials. Obliquely aligned or obliquely polished surfaces are also used to avoid reflection. An example of using an oblique alignment is shown in Figure 12.1(b). The reflected light from the obliquely aligned surface cannot input or propagate in a fiber, because the reflected light has a large propagation angle to the fiber axis. Bad splices and connectors can be considered as optical attenuators; they are shown in Figure 12.1(c,d). In Figure 12.1(d), fiber tapers are used [1]. Variable attenuation is obtained by rotating the absorption disk, whose attenuation is different for different disk placement in the case of Figure 12.1(a). Variable attenuation is also obtained by changing space s in the case of Figure 12.1(d).

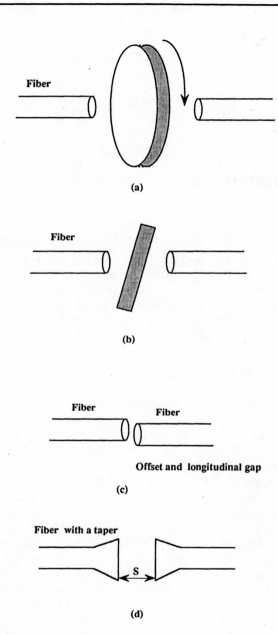

Figure 12.1 Optical attenuator.

12.2 OPTICAL ISOLATORS

Isolators are two-port optical circuits. A model is shown in Figure 12.2. The scattering matrix is

$$[S] = \begin{bmatrix} s_{11} & s_{12} \\ s_{21} & s_{22} \end{bmatrix} \tag{12.1}$$

Insertion loss L and isolation I are expressed by the scattering matrix:

$$L = 10 \log \frac{P_i}{P_t} = 10 \log \frac{1 - |S_{11}|^2}{|S_{21}|^2} \tag{12.2}$$

$$I = 10 \log \frac{|a_2|^2}{|b_1|^2} = 10 \log \frac{1}{|S_{12}|^2} \tag{12.3}$$

For an ideal isolator, $L = 0$ dB and $I = \infty$. Isolators are used in front of a laser diode to avoid reinjection of a laser light into a laser cavity, which changes or degrades the laser light's characteristics. In an EDFA application, they are used for stable performance.

The principle of an optical isolator is shown in Figure 12.3. Input light is linearly polarized by a polarizer, and then the polarization angle is rotated by a rotator with $\pi/4$ and passed through an analyzer. The reflected light propagates backward and the polarization angle is again rotated by a rotator with $\pi/4$, resulting in a passed light with polarization angle $\pi/2$. The light with polarization angle $\pi/2$ cannot pass through the polarizer, so the isolator function is realized. As for a polarizer and an analyzer, several types have been proposed and are possible, such as (1) the polarization prism (Glan-Taylor prism [2]), (2) the metal-dielectric multilayer structure (LAMIPOL [3,4]), (3) the PANDA-fiber polarizer and analyzer [5], and (4) the polarizing glass or plastic plate. Among them, (1) and (4) are well known. LAMIPOL has been proposed for the construction of a miniaturized isolator, which has a periodic metal-dielectric layer structure [3,4]. The periodic metal-dielectric

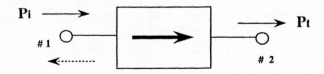

Figure 12.2 Optical isolator.

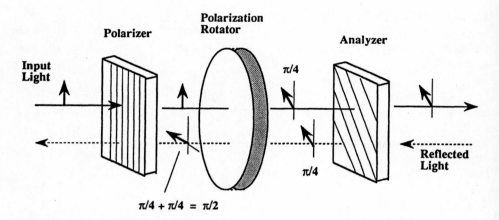

Figure 12.3 Principle of optical isolators.

layer structure serves as a polarizer, analogous to a wire grid structure in a micro-wave region. For the fabricated LAMIPOL with 20- to 30-μm thickness, the measured transmission losses of TM and TE waves are 0.6 dB and over 30 dB, respectively [4]. Generally, single-polarization characteristics in highly birefringent fibers such as a PANDA fiber can be used for a polarizer. The reported extinction ratio of the PANDA-fiber polarizer is 37 dB for a 3 m-long coiled fiber wound around the drum with a 21-cm diameter [5].

The polarization rotator is based on the Faraday effect, which is expressed as

$$\theta = VHL \qquad (12.4)$$

where θ, V, H, and L are the polarization rotating angle, the Verdet constant, the magnetic field, and the length of the sample, respectively. As shown in Figure 12.3, $\pi/4$ rotation is required. For realizing a compact isolator, length L must be small. Therefore, V and H must be large to realize $\theta = \pi/4$ with a small L value. The Verdet constant V depends on the material and the wavelength. Fiber itself (fused silica fiber) has a small V value ($V = 0.0128$ min/cm Oe at $\lambda = 0.633$ μm). For terbium-doped glass, $V = 0.25$ min/cm Oe at $\lambda = 0.633$ μm [6]. To obtain $\theta = \pi/4$ in the case of $V = 0.25$ min/cm Oe, L is about 10.8 cm for $H = 1,000$ Oe. The oersted unit has the following relationship with the ampere per meter unit:

$$1 \text{ [Oe]} = 10^3/4\pi[A/m] \qquad (12.5)$$

YIG (yttrium-iron-garnet: $Y_3Fe_5O_{12}$), which is commonly used as components of a resonator or a filter in a microwave field, is transparent at wavelength λ over 1.1 μm and has a large V value. To obtain $\theta = \pi/4$, L is about 2 mm at $\lambda = 1.3$ μm in a saturated magnetic field. Besides YIG, GdBiG (bismuth-substituted gadolinium-iron-garnet: $Gd_{3-x}Bi_xFe_5O_{12}$) and TbBiIG (bismuth-substituted terbium-iron-garnet: $Tb_{3-x}Bi_xFe_5O_{12}$) have been investigated as a Faraday rotator [7,8]. TbBiIG is reported to have good properties for constructing a wideband isolator in the 1.5-μm wavelength range [8].

Some examples of proposed isolators are shown in Figure 12.4. One uses prisms as a polarizer and an analyzer [2] and another uses PANDA fiber [5]. The reported performance of the isolator shown in Figure 12.4(a) is an insertion loss of 2 dB and isolation of 30 dB at $\lambda = 1.15$ μm [2], and that of the isolator shown in Figure 12.4(b) is an insertion loss of 4.1 dB and isolation of about 35 dB at $\lambda = 1.3$ μm [5]. These isolators use a samarium-cobalt permanent magnet for generating the magnetic field. Although the Verdet constant is small, isolators using the fiber itself as a rotator have been proposed [9,10]. The drawback of this approach is that it must use a long fiber. To overcome the small Verdet constant of a fiber-type rotator, special fibers using terbium-doped glass as a core material have been proposed and fabricated for a fiber-type rotator [6]. All these isolators have polarization-dependent characteristics. Polarization-independent characteristics are not required when isolators are used in front of a laser diode to prevent reinjection. It is known that the propagating light in an ordinary fiber (not a polarization-maintaining fiber) generally has various polarization components. Therefore, polarization-independent characteristics are necessary when they used as an inline isolator.

The polarization-independent isolator was proposed in 1979 [11]. The configuration of the proposed isolator is shown in Figure 12.5. It uses two birefringent plates with equal thickness, a Faraday rotator and a compensating plate, which rotate the polarity of light with $\pi/4$. The light with two orthogonal polarizations propagates in the directions indicated by the solid and dotted lines in the figure. The birefringent plates produce the spatial walk-off for the orthogonally polarized lights. The two orthogonally polarized lights in the case of forward light interchange their polarity before the second birefringent plate and then they are combined at the second fiber (fiber 2). In the case of the backward light, the two orthogonally polarized lights do not change their polarity because of the nonreciprocal characteristics in a Faraday rotator. Therefore, they are not combined by the function of the birefringent plate and the backward lights are not coupled into the input fiber (fiber 1). The principle used in this configuration is that the backward light focuses on the offset points at the input fiber (fiber 1). Based on this principle, a very small fiber-embedded polarization-independent isolator has been fabricated and is shown in Figure 12.6 [12]. An isolator chip is inserted between TEC fibers. The chip consists of spatial walk-off polarizers (SWP), a half-wave plate ($\lambda/2$ plate), and a Faraday rotator. Rutile plates were used as an SWP, and a garnet crystal of $(YbTbBi)_3Fe_5O_{12}$ was used as a Faraday rotator. The reported chip size is $2 \times 3 \times 2.05$ mm^3 and is very small. The fabricated isolators using this chip have a performance of about 2.5 dB of insertion loss and over 40 dB of isolation at $\lambda = 1.55$ μm [12]. Another type of polarization-independent isolator has been proposed, and it is shown in Figure 12.7

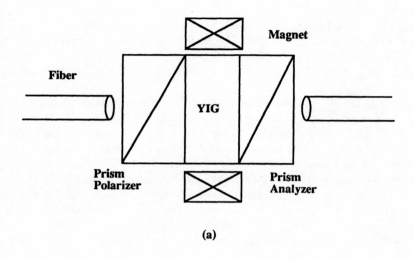

(a)

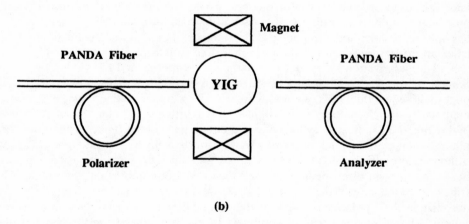

(b)

Figure 12.4 Optical isolators: (a) isolator using polarization prism (after [2]); (b) isolator using PANDA-fiber polarizer. (After [5].)

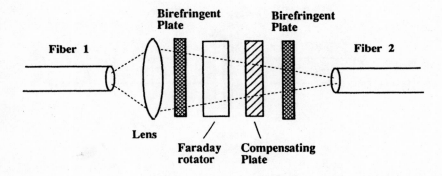

Birefringent Plate

Birefringent Plate

Fiber 1

Fiber 2

Lens

Faraday rotator

Compensating Plate

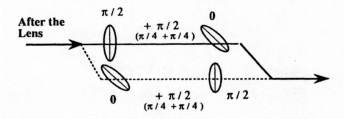

Forward Direction

After the Lens

$\pi/2$

$+\pi/2$
$(\pi/4 +\pi/4)$

0

0

$+\pi/2$
$(\pi/4 +\pi/4)$

$\pi/2$

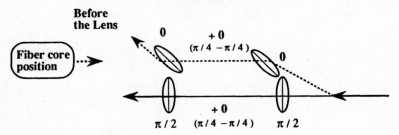

Backward Direction

Before the Lens

0

$+0$
$(\pi/4 -\pi/4)$

0

Fiber core position

$\pi/2$

$+0$
$(\pi/4 -\pi/4)$

$\pi/2$

Figure 12.5 Polarization-independent optical isolators. (After [11].)

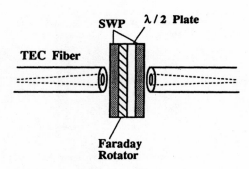

Figure 12.6 Polarization-independent optical isolators. (After [12].)

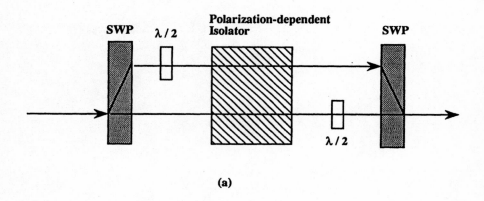

(a)

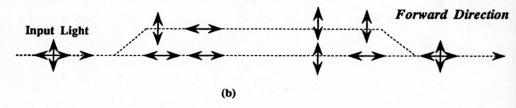

(b)

Figure 12.7 Polarization-independent optical isolators. (After [13].)

[13]. This configuration uses a polarization-dependent isolator, two SWPs, and two half-wave plates ($\lambda/2$ plates). The forward lights with orthogonal polarizations are separated by an SWP and are set in the same polarization directions by a $\lambda/2$ plate. Then they pass through the polarization-dependent isolator, and the polarization direction of one light is rotated by a $\lambda/2$ plate and two lights are combined by an SWP (Fig. 12.7(b)). The backward light is blocked by the polarization-dependent isolator. The reported performance is about 55 dB of isolation with 0.38 dB of insertion loss at $\lambda = 1.30~\mu$m, and about 64 dB of isolation with 0.32 dB of insertion loss at $\lambda = 1.55~\mu$m [13].

12.3 OPTICAL CIRCULATORS

A circulator is an n-port optical circuit. A model of a circulator is shown in Figure 12.8 for $n = 3$. In an ideal three-port circulator, input power from port #1 is output from port #2, input power from port #2 is output from port # 3, and input power from port #3 is output from port #1, and there exist no reflections for each port. The scattering matrix for $n = 3$ is expressed as

$$[S]_{\text{ideal}} = \begin{bmatrix} 0 & 0 & 1 \\ 1 & 0 & 0 \\ 0 & 1 & 0 \end{bmatrix} \qquad (12.6)$$

One application of an optical circulator has been proposed for use in the configuration of an EDFA [14,15]. Two proposed configurations using a circulator are shown

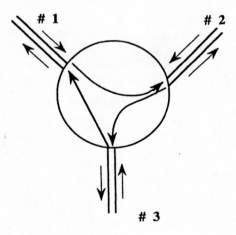

Figure 12.8 Optical circulator ($n = 3$).

in Figure 12.9 (a, b). These configurations with an optical circulator and a mirror reuse reflected pump light and also realize go/back double amplification. Therefore, a highly efficient use of the pump light and high gain amplification are possible. For these applications, not all ports in a circulator are used. Therefore, a quasicirculator, which does not satisfy the scattering matrix expressed in (12.6), is satisfactory for these applications. One example of a quasicirculator is a beam from the last port that is not output at the first port and the other port (except for the last port) works as an ordinary circulator.

There are many proposed configurations for optical circulators, and some of them are listed in [16–23]. Here, we pick up one configuration of a polarization-

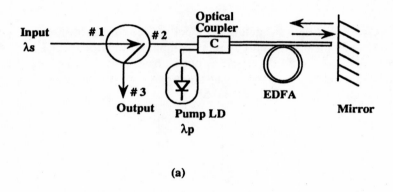

(a)

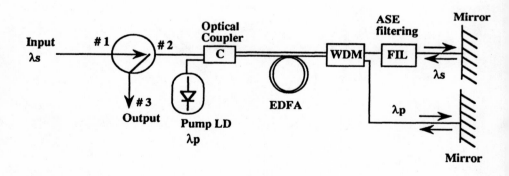

(b)

Figure 12.9 EDFA using an optical circulator. (After [14,15].)

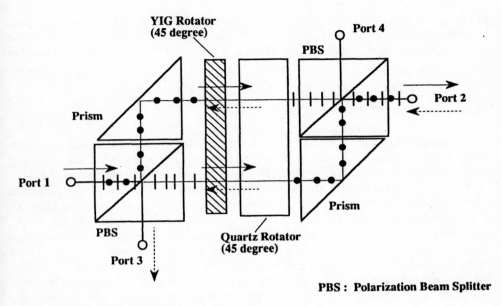

Figure 12.10 Principle of optical circulator. ([After [17].)

independent circulator to use as an example [17] to explain the working principle of an optical circulator. This configuration, which was proposed at the early investigation stage, is shown in Figure 12.10. It uses a YIG rotator, a quartz rotator, two polarization beam splitters (PBS), and two right-angle prisms. Two orthogonally polarized beams from port 1 split at the first PBS, as shown in the figure, of a solid arrow, and each beam travels separately. In the figure, two polarized beams are represented by short solid bars and small solid circles. Each beam experiences polarizing rotation and the polarization is interchanged. The interchanged-polarization beams pass through the second PBS and are output at the port 2. The input light from port 2 is separated by the PBS; however, the polarization of two separated beams is unchanged at the output of the YIG rotator because of its nonreciprocal character. Then two beams are combined at the PBS and output at the port 3. The reported dimensions of the circulator shown in Figure 12.10 are $26 \times 40 \times 15$ mm^3. The measured insertion loss and isolation are reported to be about 1.3 dB and 16 to 19 dB at a 1.32-μm wavelength, respectively [17]. Instead of using prisms and polarization beam-splitting cubes, a fiber-type circulator using fiber-type polarizers and fiber-type PBSs was proposed in [20]. The proposed polarization-independent circulator configuration uses two YIG balls, PANDA-fiber polarizers, and PBSs. The reported insertion loss and isolation are about 2.5 dB and 18 to 20 dB at a 1.30-μm wavelength, respectively [20]. A circulator using birefringent crystals (BC) instead of PBSs has been proposed [21]. The main reason for using BCs is that they have an extinction ratio that is superior to that of the PBSs. The demonstrated circulator has a performance of about 1.5 dB of insertion loss and about 42 dB of isolation [21].

12.4 FIBER-TYPE POLARIZERS

Fiber-type polarizers generally have low insertion loss, and they are useful especially when used in inline components. Polarizers are realized when only one polarized lightwave is propagated. Two types of fiber-type polarizers have been proposed. One type uses a metal coating [24–26] or a BC [27] to suppress one of the polarized lightwaves. The structure of this type is shown in Figure 12.11(a–d). In Figure 12.11(a–c), a metal coating [24–25] or a BC [27] contacts one side of the fiber with the removed cladding. In the case of using metal coating, the propagation loss difference between two modes, whose electric fields are parallel or perpendicular to the metal surface, is used for the realization of one polarized lightwave propagation. To realize the high extinction ratio with a low insertion loss, a properly designed buffer layer is introduced between the cladding and the metal surface. A single-mode fiber-type polarizer having a 45-dB extinction ratio with insertion loss less than 1 dB has been demonstrated by using aluminum and CaF_2 for the metal and buffer materials [25]. In this case, the cladding-removed fiber length L has been reported to be about 22 mm. In the case of the BC, the propagation loss difference between two polarized modes, which originates from the different field penetration characteristics of a fiber due to the BC, is used for the realization of one polarized lightwave propagation. When the refractive index of the crystal is greater than that of the fiber, the field of the fiber tends to penetrate into the crystal, resulting in a large propagation loss. On the other hand, the field of the fiber is well confined in the fiber when the refractive index of the crystal is less than that of the fiber. The refractive index of the BC is not uniform and differs for the different wave propagation angle. One polarization wave is propagated with a large radiation loss by using the proper BC, while the other is propagated with a small radiation loss. A single-mode fiber-type polarizer with an extinction ratio over 60 dB has been demonstrated by using the potassium pentaborate crystal [27]. The above-mentioned polarizers, shown in Figure 12.11(a–c), requires polishing of the cladding for them to be fabricated them. To eliminate the polishing process, the use of an eccentric core fiber, which is fabricated from an eccentric core preform, has been proposed [26]. The plastic coating in a short section (3 to 5 cm) is stripped and etched. In the etching process, the fiber is uniformly etched; however, the cladding thickness is not uniform due to the eccentric core fiber. After the etching process, a metal is sputtered or evaporated. The structure is shown in Figure 12.11(d). Based on this fabrication technique, a single-mode fiber-type polarizer having a 41-dB extinction ratio with an insertion loss of 0.31

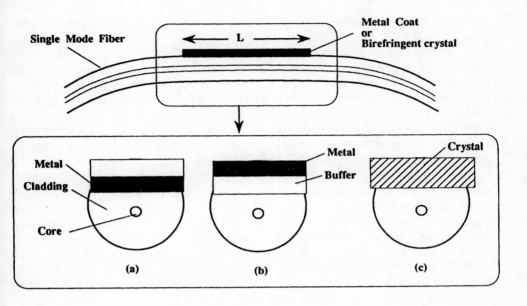

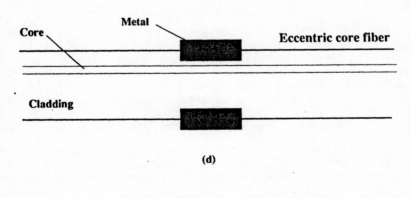

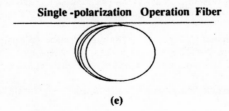

Figure 12.11 Fiber-type polarizer.

dB/cm by using aluminum for the metal and the buffer materials has been reported [26].

Another type of polarizer uses single polarization fibers or the single-polarization operation of polarization-maintaining fibers, as shown in Figure 12.11(e). In single polarization fibers, only one polarization mode of the two fundamental modes can propagate with a small propagation loss. Therefore, this type of fiber acts as a polarizer. Several types of fibers have been proposed theoretically and investigated experimentally [28–34]. Ordinary polarization-maintaining fibers, such a PANDA fiber, support two orthogonally polarized modes; however, the coupling between the two modes is very small. Therefore, the polarization-maintaining fibers preserve a polarization state over long lengths. Although they cannot be used as a polarizer in ordinary conditions, a polarizer can be constructed from them when the single-polarization operation conditions are satisfied. The single-polarization operation of polarization-maintaining fibers was investigated in [35–40]. Polarization-maintaining fibers can be properly designed so that one polarization mode propagates with a small bending loss, while another mode propagates with a large bending loss. This characteristic is wavelength-dependent. For example, an extinction ratio of 40 dB with about 0.05 dB of insertion loss at a 1.3-μm wavelength was obtained using a PANDA fiber 10m long in a 10-cm radius bend state [40].

REFERENCES

[1] Benner, A., H. M. Presby, and N. Amitay, "Low-Reflectivity In-Line Variable Attenuator Utilizing Optical Fiber Tapers," IEEE J. Lightwave Technol., Vol. 8, 1990, p. 7.

[2] Shibukawa, A., A. Katsui, H. Iwamura, and S. Hayashi, "Compact Optical Isolator for Near-Infrared Radiation," Electron. Lett., Vol. 13, 1977, p. 721.

[3] Kawakami, S., "Light Propagation Along Periodic Metal-Dielectric Layers," Appl. Opt., Vol. 22, 1983, p. 2426.

[4] Shiraishi, K., S. Sugayama, K. Baba, and S. Kawakami, "Microisolator," Appl. Opt., Vol. 25, 1986, p. 311.

[5] Okamoto, K., H. Miyazawa, J. Noda, and M. Saruwatari, "Novel Optical Isolator Consisting of a YIG Spherical Lens and PANDA-Fiber Polarisers," Electron. Lett., Vol. 21, 1985, p. 36.

[6] Shiraishi, K., S. Sugayama, and S. Kawakami, "Fiber Faraday Rotator," Appl. Opt., Vol. 23, 1984, p. 1103.

[7] Takeuchi, H., S. Ito, I. Mikami, and S. Taniguchi, "Faraday Rotation and Optical Absorption of a Single Crystal of Bismuth-Substituted Gadolium Iron Garnet," J. Appl. Phys., Vol. 44, 1973, p. 4789.

[8] Tamaki, T., H. Kaneda, and N. Kawamura, "Magneto-Optical Properties of $(TbBi)_3Fe_5O_{12}$ and Its Application to a 1.5 μm Wideband Optical Isolator," J. Appl. Phys., Vol. 70, 1991, p. 4581.

[9] Turner, E. H., and R. H. Stolen, "Fiber Faraday Circulator or Isolator," Opt. Lett., Vol. 6, 1981, p. 322.

[10] Warbrick, K., "Single-Mode Optical Isolator at 1.3 μm Using All-Fiber Components," Electron. Lett., Vol. 22, 1986, p. 711.

[11] Matsumoto, T., "Polarization-Independent Isolators for Fiber Optics," Trans. IECE of Japan, Vol. E62, 1979, p. 516; or Trans. IECE of Japan, Vol. J62-C, 1979, p. 505.

[12] Shiraishi, K., T. Chuzenji, and S. Kawakami, "Polarization-Independent In-Line Optical Isolator With Lens-Free Configuration," IEEE J. Lightwave Technol., Vol. 10, 1992, p. 1839.

[13] Shiraishi, K., "New Configuration of Polarization-Independent Isolator Using a Polarization-Dependent One," Electron. Lett., Vol. 27, 1991, p. 302.

[14] Nishi, S., K. Aida, and K. Nakagawa, "Highly Efficient Configuration of Erbium-Doped Fiber Amplifier," Proc. 16th European Conf. on Optical Communications (ECOC'90), Vol. I, 1990, p. 99.

[15] Lauridsen, V., R. Tadayoni, A. Bjarklev, J. H. Povlsen, and B. Pedersen, "Gain and Noise Performance of Fiber Amplifiers Operating in New Pump Configurations," Electron. Lett., Vol. 27, 1991, p. 327.

[16] Shibukawa, A., and M. Kobayashi, "Compact Optical Circulator for Near-Infrared Region," Electron. Lett., Vol. 14, 1978, p. 816.

[17] Iwamura, H., H. Iwasaki, K. Kubodera, Y. Torii, and J. Noda, "Simple Polarization-Independent Optical Circulator for Optical Transmission Systems," Electron. Lett., Vol. 15, 1979, p. 830.

[18] Matsumoto, T., and K. Sato, "Polarization-Independent Optical Circulator: An Experiment," Appl. Opt., Vol. 19, 1980, p. 108.

[19] Shirasaki, M., H. Kuwahara, and T. Obokata, "Compact Polarization-Independent Optical Circulator," Appl. Opt., Vol. 20, 1981, p. 2683.

[20] Yokohama, I., K. Okamoto, and J. Noda, "Polarization-Independent Optical Circulator Consisting of Two Fiber-Optic Polarising Beam Splitters and Two YIG Spherical Lenses," Electron. Lett., Vol. 22, 1986, p. 370.

[21] Koga, M., and T. Matsumoto, "Polarization-Insensitive High-Isolation Nonreciprocal Device for Optical Circulator Application," Electron. Lett., Vol. 27, 1991, p. 903.

[22] Fujii, Y., "High-Isolation Polarization-Independent Optical Circulator Coupled With Single-Mode Fibers," IEEE J. Lightwave Technol., Vol. 9, 1991, p. 456.

[23] Fujii, Y., "High-Isolation Polarization-Independent Quasi-Optical Circulator," IEEE J. Lightwave Technol., Vol. 10, 1992, p. 1226.

[24] Eickhoff, W., "In-Line Fiber-Optic Polarizer," Electron. Lett., Vol. 16, 1980, p. 762.

[25] Gruchmann, D., K. Petermann, L. Staudigel, and E. Weidel, "Fiber-Optic Polarizers With High Extinction Ratio," European Conf. on Optical Communications (ECOC'83), 1983, p. 305.

[26] Hosaka, T., K. Okamoto, and J. Noda, "Single-Mode Fiber-Type Polarizer," IEEE. J. Quantum Electron., Vol. QE-18, 1982, p. 1569.

[27] Bergh, R. A., H. C. Lefevre, and H. J. Shaw, "Single-Mode Fiber-Optic Polarizer," Opt. Lett., Vol. 5, 1980, p. 479.

[28] Okoshi, T., and K. Oyamada, "Single-Polarisation Single-Mode Optical Fiber With Refractive-Index Pots on Both Sides of the Core," Electron. Lett., Vol. 16, 1980, p. 712.

[29] Hosaka, T., K. Okamoto, Y. Sasaki, and T. Edahiro, "Single Mode Fibers With Asymmetric Refractive Index Pits on Both Sides Of Core," Electron. Lett., Vol. 17, 1981, p. 191.

[30] Okoshi, T., "Single-Polarization Single-Mode Optical Fibers," IEEE. J. Quantum Electron., Vol. QE-17, 1981, p. 879.

[31] Eickhoff, W., "Stress-Induced Single-Polarisation Single-Mode Fiber," Opt. Lett., Vol. 7, 1982, p. 629.

[32] Okoshi, T., K. Oyamada, M. Nishimura, and H. Yokota, "Side-Tunnel Fibre: An Approach to Polarisation-Maintaining Optical Waveguide Scheme," Electron. Lett., Vol. 18, 1982, p. 824.

[33] Simpson, J. R., R. H. Stolen, F. M. Sears, W. Pleibel, J. B. MacChesney, and R. E. Howard, "A Single-Polarization Fiber," IEEE J. Lightwave Technol., Vol. LT-1, 1983, p. 370.

[34] Birch, R. D., M. P. Varnham, D. N. Payne, and K. Okamoto, "Fabrication of a Stress-Guiding Optical Fibre," Electron. Lett., Vol. 19, 1983, p. 866.

[35] Snyder, A. W., and F. Ruhl, "Single Mode, Single Polarization Fibers Made of Birefringent Material," J. Opt. Soc. Am., Vol. 73, 1983, p. 1165.

[36] Snyder, A. W., and F. Ruhl, "New Single-Mode Single-Polarization Optical Fiber," Electron. Lett., Vol. 19, 1983, p. 185.

[37] Varnham, M. P., D. N. Payne, R. D. Birch, and E. J. Tarbox, "Single-Polarisation Operation of Highly Birefringent Bow-Tie Optical Fibers," Electron. Lett., Vol. 19, 1983, p. 246.

[38] Varnham, M. P., D. N. Payne, R. D. Birch, and E. J. Tarbox, "Bend Behavior of Polarising Optical Fibres," Electron. Lett., Vol. 19, 1983, p. 679.

[39] Okamoto, K., "Single-Polarization Operation in Highly Birefringent Optical Fibers," Appl. Opt., Vol. 23, 1984, p. 2638.

[40] Okamoto, K., K. Takada, M. Kawachi, and J. Noda, "All-PANDA-Fiber Gyroscope With Long-Term Stability," Electron. Lett., Vol. 20, 1984, p. 429.

Chapter 13
Optical Mechanical Switch

Optical switches (SW) have and will be used for several applications. Several types of optical switches, such as the mechanical switch, the electro-optical switch, and the laser diode switch, have been proposed and used. In this chapter, we only discuss mechanical switches that use fiber connection technologies. These switches have been used for some transmission and measurement systems.

13.1 GENERAL ASPECTS OF OPTICAL MECHANICAL SWITCHES

13.1.1 Application and Classification of Optical Switches

Optical switches have several applications in fiber-optic communication systems. Some of them are shown in Figure 13.1. Optical exchange has been investigated for broadband signal switching without conversion of optical signals into electrical signals (Fig. 13.1(a)). The switches used for an optical exchange are required to have high-speed switching characteristics and to have a potential for large-scale integration. Therefore, mechanical switches are not suitable for this application. The second example is shown in Figure 13.1(b). It is a fiber routing application for realizing reliable fiber transmission routes. This application does not require very-high-speed switching; however, wavelength-independent and polarization-independent switches are required. Mechanical switches may be used for this application. The third example is the measurement application (Fig. 13.1(c)). Optical switches are used to select a fiber in the measuring systems as shown in the figure. The switch selects one fiber among many because of the restricted amount of measurement equipment. Mechanical switches are and will be used for this application. The fourth example is the protection of equipment (Fig. 13.1(d)). When one piece of equipment is broken or out of order, another is switched on to connect to a fiber as a replacement. When the required switching time is not very short, mechanical switches may be used for this application.

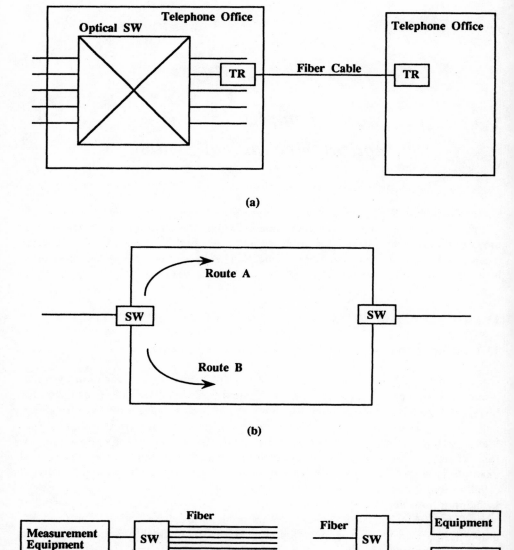

Figure 13.1 Application of optical switches: (a) optical exchange; (b) fiber routing; (c) fiber switch for measurements; (d) protection.

Optical switches are classified according to their functions. This is shown in Figure 13.2, where switches are classified into ON/OFF, I/O change ($1 \times N$), and combination change ($N \times M$) switches. The ON/OFF switch has the function of passing or blocking light. Although ON/OFF switches can be aligned or assembled in an array, this type of switch is a 1×1 switch from the functional standpoint. The $1 \times N$ switch changes an input or output (I/O) port (Fig. 13.2(b)). The $N \times M$ switch changes the combination (Fig. 13.2(c)). By combining optical switches

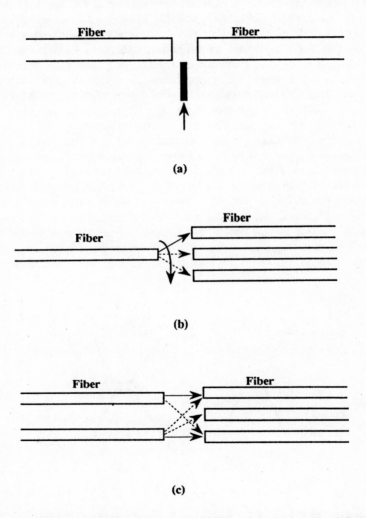

Figure 13.2 Classification of optical switches: (a) ON/OFF switch; (b) $1 \times N$ switch (changes I/O); (c) $N \times M$ switch (changes combination).

and couplers (directional couplers or star couplers), several functions, which are not shown in Figure 13.2, may be realized. For example, some of the N ports are switched to some of the M ports (here, we tentatively name this group switching). When we consider several switches from the standpoint of a basic optical switching unit, three classification types may be sufficient. Other classifications are possible from several standpoints. From the working principle, optical switches are classified into mechanical and nonmechanical switches. For a nonmechanical switch type, there are waveguide-type and amplifier-type switches. Semiconductor waveguide switches, silica-based waveguide switches, and LiNbO$_3$ switches belong to the waveguide-type switch. The working principle of an amplifier-type switch is that light passes with amplification or no loss for the normal amplification state, and light is blocked for the nonamplification state such as a shutdown of the driving current. A laser diode switch and an EDFA switch belong to this amplifier-type switch. The definition of a mechanical switch is that it uses some mechanical movements for switching. Mechanical switches are classified into two types: (1) fiber-moving type and (2) optical-component-moving type. Connector-moving or ferrule-moving type switches are included in the first type, which has the advantages of low insertion loss because no other components are inserted between fibers. Prisms, mirrors, and lenses are used for optical components in the second type of mechanical switches. Since mechanical switches are precisely explained according to this classification in the following sections, some brief explanations for other types of switches are provided here by using Figure 13.3. In Figure 13.3(a), a Mach-Zehnder interferometer with a phase shifter is formed in a waveguide-type optical circuit, such as silica-based or semiconductor-based planar circuits. As can be understood from previous chapters, this configuration works as an optical ON/OFF switch by changing the phase of the phase shif-

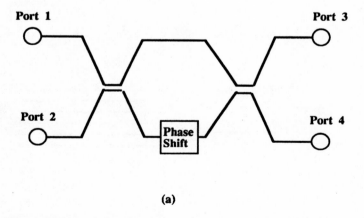

(a)

Figure 13.3 Several optical switches: (a) MZ-type ON/OFF switch; (b) 1 × 2 switch.

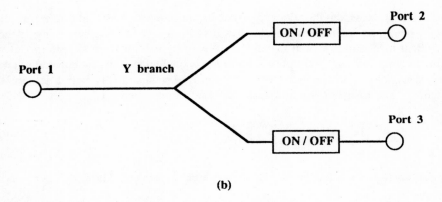

(b)

Figure 13.3 (Continued)

ter. The fabricated silica-based planar circuit has been reported in [1] and the phase shift is accomplished by a thermo-optical effect using a thin-film heater. The second example is an optical $1 \times N$ switch ($N = 2$ in the case of Fig. 13.3(b)), which uses a star coupler (a Y-branch) and ON/OFF gate switches. One example of ON/OFF gate switches is a laser diode switch [2].

13.1.2 Characteristics of Optical Mechanical Switches

The characteristics of optical mechanical switches are listed in Table 13.1 as they compare to other types of switches. The advantages of mechanical switches are: low

Table 13.1
Characteristics of Optical Mechanical Switch

	Mechanical SW	*Semiconductor or Electro-optical SW*
Size	Large	Small
Large-scale integration	Difficult	May be easy
Switching speed	Slow	Fast
Insertion loss	Low (<0.5 dB)	Not very low
Wavelength dependence	Independent for fiber transmission	Dependent (usually)
Bit rate bandwidth	Bit rate–free; wide bandwidth	Relatively wide bandwidth
Polarization dependence	Independent	Dependent (usually)

insertion loss, wavelength independence, bit rate independence, and polarization independence. The main drawback is slow switching speed. The switching speed of the optical mechanical switch is considered here by using a very simple model with simple assumptions. As a simple model of a mechanical switch, we consider a switch where some object with a mass m must be moved mechanically with a length L for switching. The energy E for moving with a constant speed v is

$$E = \frac{1}{2} m v^2 \qquad (13.1)$$

and this energy is assumed to be fed by an electric energy. Then the energy is

$$E = pt \qquad (13.2)$$

where p and t are the electric power and the power feeding time, respectively. We assume that the power feeding time t is equal to the switching time. This means that the electric power accumulation is not assumed in this model; this assumption may be held for a switch using magnetic coils for driving. With this simple assumption, the switching time t is expressed by

$$t = \frac{L}{v} = \frac{L\sqrt{m}}{\sqrt{2pt}} \qquad (13.3)$$

Although the model is not realistic, a rough estimation may be possible. This equation is derived by using (13.1) and (13.2) and no energy loss is assumed. By solving (13.3), switching time t is obtained as

$$t = \frac{L^{2/3} m^{1/3}}{(2p)^{1/3}} \qquad (13.4)$$

For example, we use $m = 10$ g, $L = 250$ μm, and $p = 10W$. Then $t = 0.3$ ms. This example shows that the switching time of a simple model is on the order of a submillisecond and is not enough for high-speed switching applications.

13.2 1 × 1 MECHANICAL SWITCH (ON/OFF SWITCH)

Several ON/OFF mechanical switches are possible, as shown in Figure 13.4. The switch in Figure 13.4(a) belongs to the fiber-moving type and that in Figure 13.4(b–d) belongs to the optical-component-moving type. In the case of a fiber-moving type, one fiber or a ferrule is moved manually or by some electromagnetic force or other

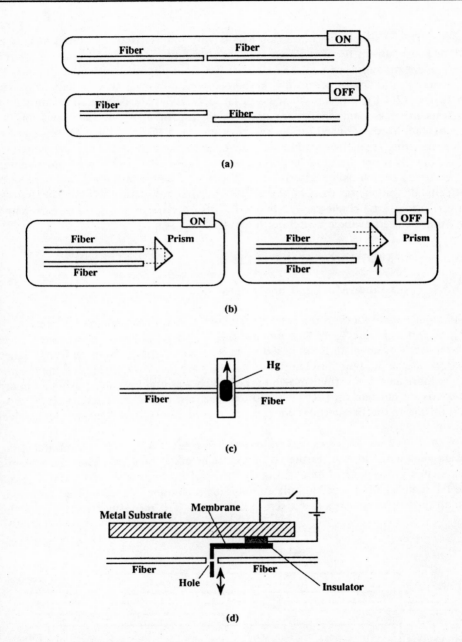

Figure 13.4 ON/OFF switches: (a) fiber-moving switch; (b) prism-moving switch; (c) electrowetting switch (based on [3]); (d) electrostatic switch. (After [4].)

means. This is simple, and insertion loss is very low. However, a repeatable finely aligned mechanism for two fibers must be developed. On the other hand, optical-component-moving type switches are easy for fiber alignment because fibers are in stationary states. Insertion loss for optical-component-moving types is relatively high. In Figure 13.4(b), a prism is used and an ON/OFF state is created by moving it. Mirrors or other components are possible instead of a prism. Electrowetting and electrostatic force are used for the switches shown in Figure 13.4(c,d), respectively. Electrowetting is an electrically induced wettability change. Although the switch discussed in [3] is not the same as that shown in Figure 13.4(c), the reported switching time for the switch using this effect is 20 ms. A membrane with a hole is electrostatically driven in the case of the switch shown in Figure 13.4(d). The reported insertion loss for a single-mode fiber is 1.2 dB on average at a 1.3-μm wavelength, and the reported average switching time is 1.2 ms [4].

13.3 1 × *N* MECHANICAL SWITCH

13.3.1 1 × 2 Fiber-Moving-Type Switch

The simple configuration for the 1 × 2 fiber-moving-type switch was proposed in [5], shown in Figure 13.5. One moving fiber is mated to either of two fixed fibers. The moving mechanism is supposed to be a magnetic force acting on ferromagnetic sleeves coated onto the moving fiber. It is only the proposal in [5], but one example of the fabricated 1 × 2 fiber-moving-type switch was reported in [6] using a bimorph as a driving mechanism [6]. The reported insertion loss and switching time are 0.7 dB, including the Fresnel loss for 60-μm core SI fibers, and 0.84 ms, respectively [6].

A 1 × 2 switch using plastic-molded ferrules has been proposed and it uses rectangular holes in one ferrule and cylindrical guide pins [7]. The mechanism of this switch is shown in Figure 13.6. Both ferrules, a movable and a fixed ferrule, are fabricated with a precision plastic-molding technique, which is used for making the array-fiber connectors discussed in Chapter 8. The movable ferrule with a fiber

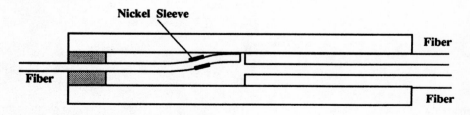

Figure 13.5 A 1 × 2 switch. (After [5].)

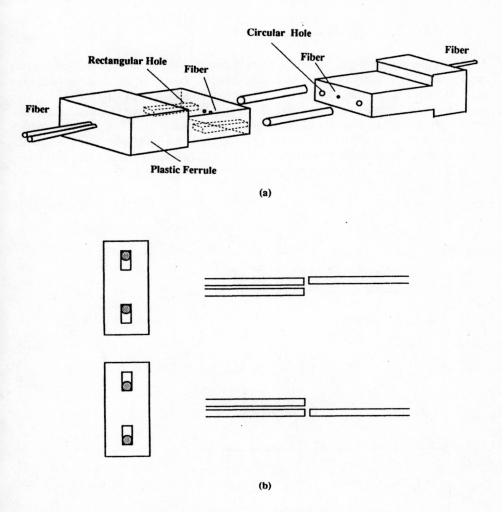

Figure 13.6 A 1 × 2 switch: (a) configuration; (b) two steady positions. (After [7].)

has two circular holes with a small clearance to a guide pin. The fixed ferrule with two fibers has two rectangular holes with a large clearance to create two steady positions for a guide pin. Switching is accomplished by moving the movable ferrule to change its position, as shown in the figure. By using an electromagnet, a switching time of 4 ms has been reported, with an average insertion loss of 0.7 dB, including a Fresnel reflection loss (0.3 dB), for 50-μm core GI multimode fibers [7].

13.3.2 1 × *N* Fiber-Moving-Type Switch

A 1 × 9 fiber-moving-type switch using a *V*-groove has been fabricated [8], and its configuration is shown in Figure 13.7. The reported insertion loss and the switching time are about 0.3 dB and less than 300 ms for 60-μm core SI fibers [8].

A 1 × *N* fiber-moving-type switch using silicon *V*-groove chips has been fabricated [9]. Figure 13.8 shows the configuration of a fabricated 1 × 8 switch, which has three blocks of the basic 1 × 2 switch. The first block has one fiber set in a 12-*V*-groove chip for input and two fibers for output. The second block has two fibers set in a 12-*V*-groove chip for input and four fibers for output. The third block has

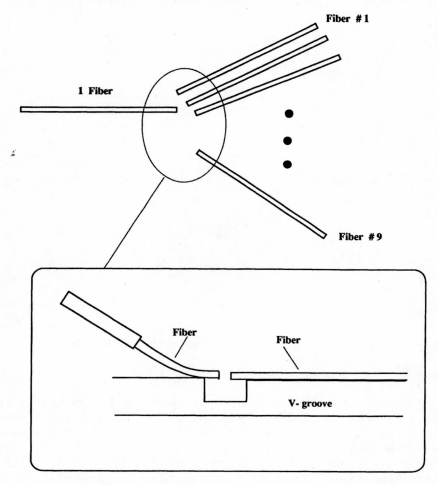

Figure 13.7 A 1 × 9 switch. (After [8].)

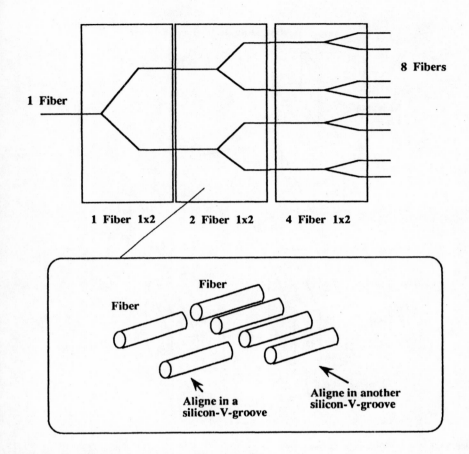

Figure 13.8 Configuration of 1 × 8 switch. (After [9].)

four fibers for input and eight fibers for output. Each block is moved by an actuator. The reported performance of these multistage moving-fiber array switches is 0.2 and 0.8 dB for 1 × 8 multimode fibers and 1 × 4 single-mode fibers, respectively.

A 1 × *N* fiber-moving-type switch can be developed using array-fiber connectors. A 1 × 50 switch for ribbons with 10 single-mode fiber has been reported [10]. The switch employs small 10-fiber ribbon connectors aligned by two guide pins, which were explained in Chapter 8. In this switch, the pin tops are tapered for the smooth insertion of the pins into the guide holes. Figure 13.9 shows the mechanism of a 1 × 50 switch. The switch has one movable plug and 50 fixed plugs that are linearly arranged in opposite locations. The movable plug is roughly transported to a desired plug position, and its then precisely aligned by inserting guide pins. The switching time of this switch is the sum of the traveling time of a moving plug, the

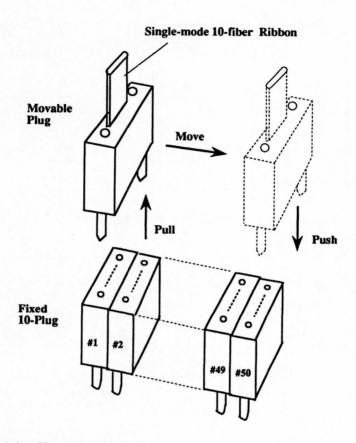

Figure 13.9 A 1 × 50 switch. (After [10].)

plug connection time, and the driving motor delay time. The plug connection time is a plug pulling and pushing time. The reported switching time from ports 1 to 50 and from ports 1 to 2 is 80 ms and 17 ms, respectively [10]. The reported insertion loss using an index-matching material is 0.47 dB on average for a single-mode 10-fiber ribbon at a 1.3-μm wavelength [10]. This switch is applicable for a 1 × N (N = 1, 2,3, ...) switch for an n-fiber ribbon (n = 1, 2,3, ...), such as a 1 × 2 switch for an eight-fiber ribbon.

13.3.3 1 × 2 Optical-Component-Moving-Type Switch

Fibers are fixed in an optical-component-moving-type switch. Prisms, mirrors, or other components are used for moving components in this mechanical switch. Some examples are shown in Figure 13.10(a,b). The first example, Figure 13.10(a), uses

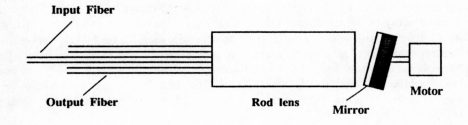

(a)

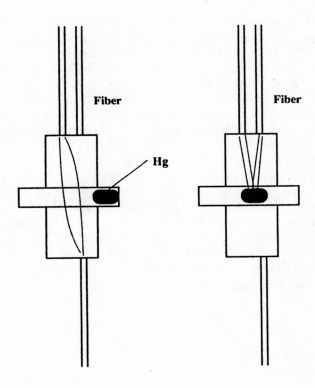

(b)

Figure 13.10 A 1 × *N* switch. (After [3,11].)

a mirror and a rod lens. A mirror is rotated to scan an image of the input fiber over a circular array of output fibers [11]. The reported insertion losses of the switch are 4 to 11 dB for 55-μm core GI fibers. The second example, Figure 10.13(b), uses mercury, which acts as a mirror, and it is moved by the electrowetting effect [3]. The reported insertion losses of this switch are about 0.5 to 2 dB for 50-μm core fibers, and the switching time is about 20 ms. These optical-component-moving-type switches have a large insertion loss, even when used for multimode fibers, compared to fiber-moving-type switches.

13.4 $N \times M$ MECHANICAL SWITCH (MATRIX SWITCH)

13.4.1 $N \times M$ Fiber-Moving-Type Switch

A 2 × 2 switch for two fibers has been fabricated using the same technology as that of the switch shown in Figure 13.6. The configuration of the 2 × 2 switch is shown in Figure 13.11 [12]. The fixed ferrule with six fibers has two rectangular holes with a large clearance to create two steady positions (normal and transferred stable positions) using guide pins. Switching is accomplished by moving the movable ferrule to change its position, as shown in the figure. This 2 × 2 switch for two fibers is realized by using fibers that are configured in a loop, as shown in Figure 13.11. By improving the accuracy of ferrule dimensions, an average insertion loss of 0.35 dB for DSFs has been reported with a switching time of 3 ms [12].

A 100 × 100 switch was fabricated in [13]. It uses optical connectors (ferrules and sleeves) and small robot hands. The robot hands disconnect and connect optical connectors. This optical matrix switch is a nonblocking type, and its configuration is shown in Figure 13.12. The ferrules on one side can move in the matrix row direction, and those on another side in the column direction. Each ferrule can be independently and arbitrarily positioned using this method. This results in an arbitrary connection between two ferrule-terminated fiber groups. The reported insertion loss and switching time for ordinary single-mode fibers are about 1 dB and less than 1.3 minutes, respectively [13].

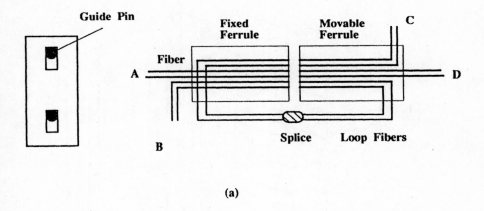

(a)

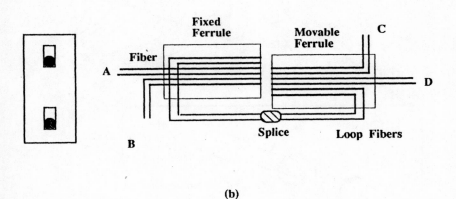

(b)

Figure 13.11 A 2 × 2 switch: (a) normal position (A-D, B-C); (b) transferred position (A-C, B-D). (After [12].)

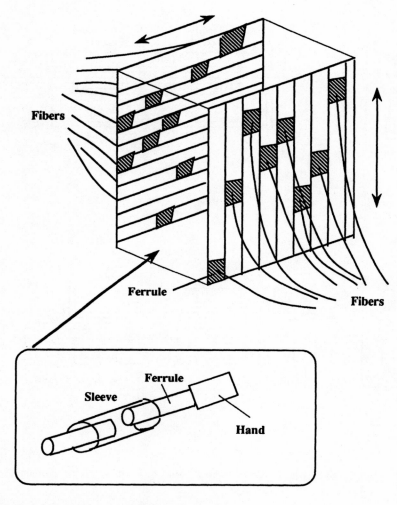

Figure 13.12 A 100 × 100 matrix switch. (After [13].)

13.4.2 *N × M* Optical-Component-Moving-Type Switch

A 4 × 4 switch has been reported, which uses a rhombic glass-block switch element [14]. The switching is done by moving the glass block as shown in Figure 13.13. The reported insertion loss and switching time are 1.3 dB (maximum) for 60-μm core GI fibers and less than 18 ms, respectively [14].

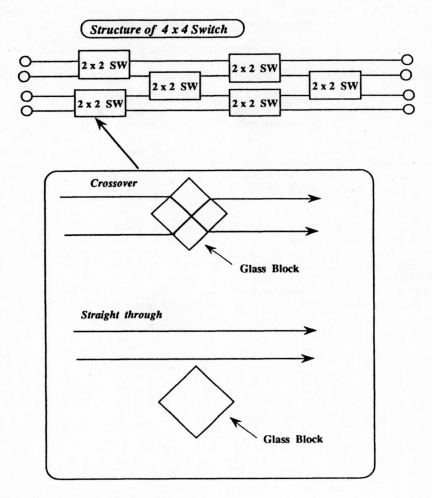

Figure 13.13 A 4 × 4 switch. (After [14].)

REFERENCES

[1] Takato, N., K. Jinguji, M. Yasu, H. Toba, and M. Kawachi, "Silica-Based Single-Mode Wave-guides on Silicon and Their Application to Guided-Wave Optical Interferometers," IEEE J. Light-wave Technol., Vol. 6, 1988, p. 1003.

[2] Ikeda, M., "Laser Diode Switch," Electron. Lett., Vol. 17, 1981, p. 899.

[3] Jackel, J. L., S. Hackwood, J. J. Veselka, and G. Beni, "Electrowetting Switch for Multimode Optical Fibers," Appl. Opt., Vol. 22, 1983, p. 1765.

[4] Hogari, K., and T. Matsumoto, "Electrostatically Driven Fiber-Optic Micromechanical On/Off Switch and Its Application To Subscriber Transmission Systems," IEEE J. Lightwave Technol., Vol. 8, 1990, p. 722.

[5] Hale, P. G., and R. Kompfner, "Mechanical Optical-Fiber Switch," Electron. Lett., Vol. 12, 1976, p. 388.

[6] Ohmori, Y., and H. Ogiwara, "Optical Fiber Switch Driven by PZT Bimorph," Appl. Opt., Vol. 17, 1978, p. 3531.

[7] Nagasawa, S., H. Furukawa, T. Satake, and N. Kashima, "A New Type of Optical Switch With a Plastic-Molded Ferrule," IEICE of Japan, Vol. E 70, 1987, p. 696.

[8] Yamamoto, H., and H. Ogiwara, "Moving Optical-Fiber Switch Experiment," Appl. Opt., Vol. 17, 1978, p. 3675.

[9] Young, W. C., and L. Curtis, "Cascaded Multipole Switches for Single-Mode and Multimode Optical Fibers," Electron. Lett., Vol. 17, 1981, p. 571.

[10] Satake, T., S. Nagasawa, and N. Kashima, "Single-Mode 1 × 50 Switch for 10-Fiber Ribbon," IEICE of Japan, Vol. E 70, 1987, p. 623.

[11] Tomlinson, W. J., R. E. Wagner, A. R. Strnad, and F. A. Dunn, "Multiposition Optical-Fiber Switch," Electron. Lett., Vol. 15, 1979, p. 192.

[12] Nagasawa, S., H. Kobayashi, and F. Ashiya, "A Low-Loss Multifiber Mechanical Switch for 1. 55 μm Zero-Dispersion Fibers," IEICE of Japan, Vol. E 73, 1990, p. 1147.

[13] Katagiri, T., Y. Koyamada, M. Tachikura, and Y. Katsuyama, "Nonblocking 100 × 100 Opto-mechanical Matrix Switch for Subscriber Networks," 40th International Wire and Cable Symp. (IWCS), 1991, p. 285.

[14] Fujii, Y., J. Minowa, T. Aoyama, and K. Doi, "Low-Loss 4 × 4 Optical Matrix Switch for Fiber-Optic Communication," Electron. Lett., Vol. 15, 1979, p. 427.

Chapter 14
Future Technology

Future technologies of passive optical components from the author's point of view are discussed in this chapter. A truly revolutionary technology is difficult to anticipate, so future technologies discussed in this chapter will be based on technology trends and analogies in electric components and devices.

14.1 FIBER AND CABLE CONNECTION

Optical-fiber manufacturing technology has made rapid progress since 1970. For example, fiber loss decreased from 20 dB/km in 1970 to about 0.2 dB/km in 1979 [1,2]. Along with the decrease in loss, fiber dimensional control technology has also improved, which makes it possible to use the fixed methods for fiber alignment in fusion splicing, even for single-mode fibers. Thanks to this improvement, low-loss mass-fusion splicing and no-adjustment-type connectors for single-mode fibers are realized today. It is reasonable to think that the fixed or no-adjustment method can be applied to most of the splices and connectors in the future, due to further improvement of fiber dimensional control technology.

Historically speaking, one-by-one splicing or connecting (individual splicing or connecting) was realized at first. Later, mass-fusion splicing and array-fiber connectors were developed and are used today. Two-dimensional connectors were proposed and fabricated at the laboratory level in 1987 and 1988 [3,4]. The connection loss of these connectors is large when compared to single- or array-fiber connectors currently. Connection loss reduction in 2D connectors must be realized in the future, because mass deployment of optical fibers in the future will require these connectors. Possible applications of 2D connectors are shown in Figures 14.1 and 14.2. The first application is optical cable connection. Today's optical cables are connected in two steps: the first step is fiber splicing or connecting, and the second step is connecting cable using a box or a closure. By using a preconnected cable with 2D connectors, these two steps are reduced to one simple fiber/cable connection step. Cables with

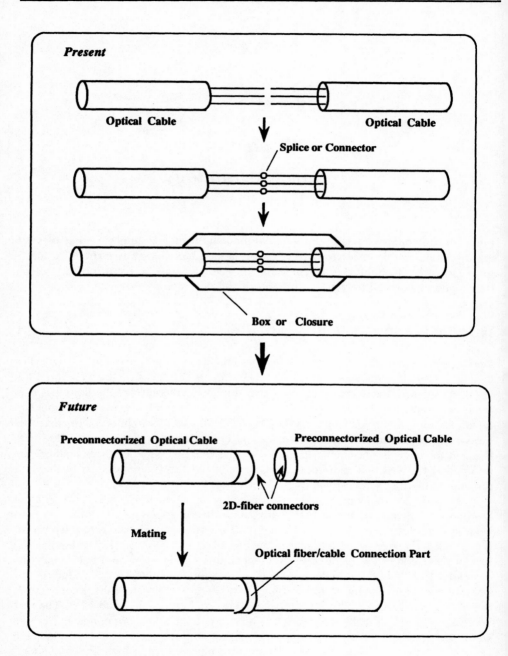

Figure 14.1 Application of 2D-fiber connectors.

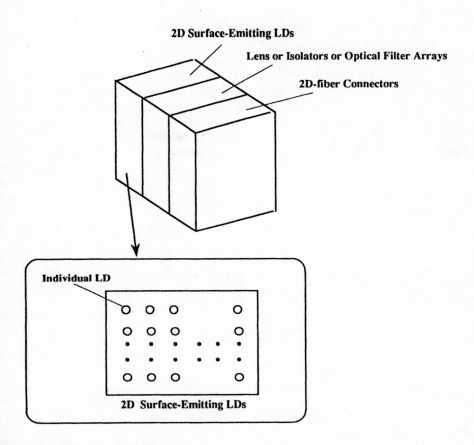

Figure 14.2 Application of 2D-fiber connectors.

a high fiber count, say 1,000-fiber cables, will be able to be connected in a minute. The second application is the compact and simple coupling of laser diodes to cables with a high fiber count. A surface-emitting laser, in which light output is taken vertically from the surface, has been proposed and fabricated [5,6], and a 2D array of surface-emitting lasers has been demonstrated [7]. By combining the 2D surface-emitting laser diode array with the 2D connectors, which are attached to cables with a high fiber count, a compact and simple coupling may be realized. In addition to application in optical transmission systems, a 2D surface-emitting laser diode array and 2D connectors will be used in the future for the applications of optical signal processing and optical interconnection.

14.2 OPTICAL SIGNAL PROCESSING AND OPTICAL INTERCONNECTION

Today's fiber-optic transmission systems use electric circuits for signal processing after the conversion of optical signals to electric signals. Compact, reliable and inexpensive lightwave circuits, if realized, have the advantage of wide bandwidth over the existing electric circuits. Like electric circuits, several passive optical components such as an optical filter and an optical delay line will be used in lightwave circuits in the future. These passive optical components will be integrated for optical signal processing. The compact and efficient coupling between the integrated lightwave circuits and optical fibers will be an important issue (Fig. 14.3(a)). Optical interconnection between these lightwave circuits or lightwave circuit boards must be developed in the future, just like the electric circuit boards. Optical signals will transmit in a fiber or in a space for optical interconnection (Fig. 14.3(b)). These two

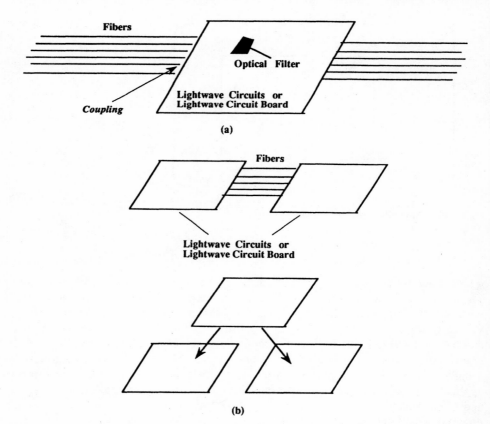

Figure 14.3 Optical interconnection.

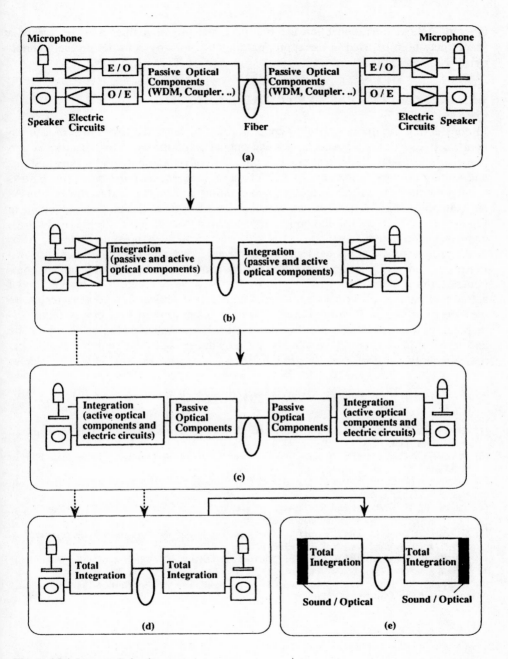

Figure 14.4 Integrated circuits.

types of optical interconnection technologies, the use of a fiber and the use of a space, may both be used in the appropriate applications by considering the technologies' characteristics.

14.3 INTEGRATED PASSIVE OPTICAL CIRCUITS

In considering the many optical systems at present, there are four types of components: active optical components, passive optical components, active electrical components, and passive electrical components. To realize compact and robust systems and system cost reduction, integration of these components must be made. Efforts to integrate have been made, and total integration will be realized in the future. As an example of the integration, we consider here the transmission system shown in Figure 14.4. When we use the optical transmission system shown in Figure 14.4(a), communication between two persons at separated locations is possible by voice. Most of the present practical optical transmission systems are similar to the system shown in Figure 14.4(a). By integrating active optical components with passive optical components, the system shown in Figure 14.4(b) will be realized. The integration of active and passive electrical components with active optical components results in the system shown in Figure 14.4(c). The total integration of four components corresponds to the system shown in Figure 14.4(d). If we realize the optical microphone and speaker in the future, which are the transducers between sound pressure and optical energy, another total integration, shown in Figure 14.4(e), may be possible.

REFERENCES

[1] Kapron, F. P., D. B. Keck, and R. D. Maurer, "Radiation Losses in Glass Optical Waveguides," Appl. Phys. Lett., Vol. 17, 1970, p. 423.
[2] Miya, T., Y. Terunuma, T. Hosaka, and T. Miyashita, "Ultimate Low-Loss Single-Mode Fiber at 1. 55 μm," Electron. Lett., Vol. 15, 1979, p. 106.
[3] Satake, T., N. Kashima, and M. Oki, "Ultra High Density 50-Fiber Connector," IEICE of Japan, Vol. E70, 1987, p. 621.
[4] Satake, T., N. Kashima, and S. Nagasawa, "Plastic Molded Single-Mode 50-Fiber Connectors," Conf. on Optical Fiber Communication (OFC'88), THJ2, 1988.
[5] Soda, H., K. Iga, C. Kitahara, and Y. Suematsu, "GaInAsP/InP Surface-Emitting Injection Lasers," Jap. J. Appl. Phys., Vol. 18, 1979, p. 2329.
[6] Iga, K., S. Kinoshita, and F. Koyama, "Microcavity GaAlAs/GaAs Surface-Emitting Laser With $I_{th} = 6$ mA," Electron. Lett., Vol. 23, 1987, p. 134.
[7] Uchiyama, S., and K. Iga, "Two-Dimensional Array of GaInAsP/InP Surface-Emitting Lasers," Electron. Lett., Vol. 21, 1985, p. 162.

About the Author

Norio Kashima was born in Hiroshima, Japan, in 1950. He received his BE and ME degrees in electrical engineering from the Yokohama National University, Yokohama, Japan, in 1973 and 1975, respectively, and a PhD in physical electronics from the Tokyo Institute of Technology, Tokyo, Japan, in 1984.

Dr. Kashima joined the Electrical Communication Laboratory of NTT in 1975, and has been engaged in the research and development of optical-fiber structural design, fusion splicing, optical connectors, subscriber loop systems, and operation systems of optical cable systems.

He is a member of the IEEE and the IEICE of Japan. He received the Young Engineer Award from IEICE of Japan in 1982. He is also the author of *Optical Transmission for the Subscriber Loop*, published by Artech House.

Index

The Artech House Antenna Library

Helmut E. Schrank, *Series Editor*

Advanced Technology in Satellite Communication Antennas: Electrical and Mechanical Design, Takashi Kitsuregawa

Analysis of Wire Antennas and Scatterers: Software and User's Manual, A. R. Djordjevic, M. B. Bazdar, G. M. Bazdar, G. M. Vitosevic, T. K. Sarkar, and R. F. Harrington

Analysis Methods for Electromagnetic Wave Problems, E. Yamashita, editor

Antenna-Based Signal Processing Techniques for Radar Systems, Alfonso Farina

CAD for Linear and Planar Antenna Arrays of Various Radiating Elements: Software and User's Manual, Miodrag Mikavica and Aleksandar Nešić

The CG-FFT Method: Application of Signal Processing Techniques to Electromagnetics, Manuel F. Cátedra, Rafael P. Torres, José Basterrechea, Emilio Gago

Electromagnetic Waves in Chiral and Bi-Isotropic Media, I.V. Lindell, S.A. Tretyakov, A.H. Sihvola, A. J. Viitanen

Fixed and Mobile Terminal Antennas, A. Kumar

Generalized Multipole Technique for Computational Electromagnetics, Cristian Hafner

Integral Equation Methods for Electromagnetics, N. Morita, N. Kumagai, and J. Mautz

IONOPROP: Ionospheric Propagation Assessment Program, Version 1.1: Software and User's Manual, by Hernert V. Hitney

Four-Armed Spiral Antennas, Robert G. Corzine and Joseph A. Mosko

Introduction to Electromagnetic Wave Propagation, Paul Rohan

Introduction to the Uniform Geometrical Theory of Diffraction, D. A. McNamara

Microwave Cavity Antennas, A. Kumar and H. D. Hristov

Millimeter-Wave Microstrip and Printed Circuit Antennas, Prakash Bhartia

Mobile Antenna Systems, K. Fujimoto and J. R. James

Modern Methods of Reflector Antenna Analysis and Design, Craig Scott

Moment Methods in Antennas and Scattering, Robert C. Hansen, editor

Monopole Elements on Circular Ground Planes, M. M. Weiner *et al.*

Near-Field Antenna Measurements, D. Slater

Passive Optical Components for Optical Fiber Transmission, Norio Kashima

Phased Array Antenna Handbook, Robert J. Mailloux

For further information on these and other Artech House titles, contact:

Artech House
685 Canton Street
Norwood, MA 01602
617-769-9750
Fax: 617-769-6334
Telex: 951-659
email: artech@world.std.com

Artech House
6 Buckingham Gate
London SW1E6JP England
+44 (0) 71-973-8077
Fax: +44 (0) 71-630-0166
Telex: 951-659
email: bookco@artech.demon.co.uk